Climate Change and Decarbonization 2018

Transportation Research Record: Journal of the Transportation Research Board

Volume 2672, Issue 2

ISBN: 978-1-5108-7730-6

Printed from e-media with permission by:

Curran Associates, Inc.
57 Morehouse Lane
Red Hook, NY 12571

Some format issues inherent in the e-media version may also appear in this print version.

Published by Sage Publications USA

ISBN (Print) 978-1-5108-7730-6
ISBN 2018 Printed Set (All Issues) 978-1-5108-7735-1

Printed by Curran Associates, Inc. (2018)

Additional copies of this publication are available from:

Curran Associates, Inc.
57 Morehouse Lane
Red Hook, NY 12571 USA
Phone: 845-758-0400
Fax: 845-758-2633
Email: curran@proceedings.com
Web: www.proceedings.com

Volume 2672 Number 2 December 2018

TRR
JOURNAL OF THE TRANSPORTATION RESEARCH BOARD

Contents

Climate Change and Decarbonization 2018

Articles

Article

TRR
JOURNAL OF THE TRANSPORTATION RESEARCH BOARD

Transportation Research Record
2018, Vol. 2672(2) 1–10

DOI: 10.1177/0361198118756366
journals.sagepub.com/home/trr
SAGE

Recent and Future Outlooks for Nuisance Flooding Impacts on Roadways on the U.S. East Coast

Jennifer M. Jacobs[1], Lia R. Cattaneo[2], William Sweet[3], and Theodore Mansfield[4]

Abstract

Tidal floods (i.e., "nuisance" flooding) are occurring more often during seasonal high tides or minor wind events, and the frequency is expected to increase dramatically in the coming decades. During these flood events, coastal communities' roads are often impassable or difficult to pass, thus impacting routine transport needs. This study identifies vulnerable roads and quantifies the risk from nuisance flooding in the Eastern United States by combining public road information from the Federal Highway Administration's Highway Performance Monitoring System with flood frequency maps, tidal gauge historic observations, and future projections of annual minor tidal flood frequencies and durations. The results indicate that tidal nuisance flooding across the East Coast threatens 7508 miles (12,083 km) of roadways including over 400 miles (644 km) of interstate roadways. From 1996–2005 to 2006–2015, there was a 90% average increase in nuisance floods. With sea level rise, nuisance-flood frequency is projected to grow at all locations assessed. The total induced vehicle-hours of delay due to nuisance flooding currently exceed 100 million hours annually. Nearly 160 million vehicle-hours of delay across the East Coast by 2020 (85% increase from 2010); 1.2 billion vehicle-hours by 2060 (126% increase from 2010); and 3.4 billion vehicle-hours by 2100 (392% increase from 2010) are projected under an intermediate low sea-level-rise scenario. By 2056–2065, nuisance flooding could occur almost daily at sites in Connecticut, New Jersey, Maryland, the District of Columbia, North Carolina, and Florida under an intermediate sea-level-rise scenario.

Transportation infrastructure in coastal regions is vulnerable to coastal extreme events today; this vulnerability will increase with sea level rise (SLR), enhanced storm surge from tropical and nontropical storms, and land subsidence (*1,2*). Hurricanes have caused billions of dollars in direct damage to coastal roadways and bridges. Significant economic losses have also occurred due to transport disruption during and after these storms. Currently, 60,000 miles (96,561 km) of roadways are exposed to coastal storms, as noted in Douglass et al. (*2*). In the future, rising seas will cause more frequent disruptions, more damage, and storm surge impacts to extend further inland.

In some coastal areas, impacts are not limited to storm events. Critical transportation infrastructure is at risk from SLR alone. SLR-induced coastal flooding decreases service and increases maintenance costs for existing facilities; mitigation strategies will be an increasingly critical aspect of transportation planning, design, and operations and maintenance (*3–5*). Numerous transportation agencies have identified assets vulnerable to permanent inundation for sea-level-rise scenarios for the United States that range from an increase in 1–8.2 feet (0.305–2.5 m) by the year 2100 with sea-level-rise values along the U.S. Atlantic and Gulf Coasts likely to be greater than the global average (*6*).

SLR is also contributing to an increasing frequency of tidal floods (i.e., "sunny day," "recurrent," "shallow coastal," and "nuisance" flooding), occurring more often during seasonal high tides or minor wind events. Some portions of the U.S. coast, including much of the Atlantic coast, are seeing an accelerating frequency of such events (*7, 8*). Flooding that surpasses local emergency thresholds for minor tidal flooding (i.e., "nuisance" levels of about 30–60 cm (1–2 ft)) result

[1]Department of Civil and Environmental Engineering, University of New Hampshire, Durham, NH
[2]Office of Policy Development, Strategic Planning and Performance, US Department of Transportation, Washington, D.C.
[3]National Ocean Service, National Oceanic and Atmospheric Administration, Silver Spring, MD
[4]Oak Ridge Institute for Science and Education, Office of Policy Development, Strategic Planning and Performance, US Department of Transportation, Washington, D.C.

Corresponding Author:
Address correspondence to Jennifer M. Jacobs: jennifer.jacobs@unh.edu

in flooded infrastructure including roadways. These nuisance floods have increased 5- to 10-fold or more since the 1960s along the U.S. coastlines (*8–12*). For example, cities such as Annapolis (Maryland), Norfolk (Virginia), and Miami Beach (Florida), are now flooded numerous times per year (42, 11, and 17 days in 2016, respectively) (*13*). With rising sea levels, such flooding is expected to increase dramatically in the coming decades (*8, 10, 14–16*).

The damage to coastal infrastructure from recurrent flooding is less well understood than that due to extreme coastal events or gradual SLR. In some regions, the cumulative cost of nuisance flooding could be comparable to, or exceed costs from, extreme coastal storm events (*9*). Dahl et al. determined that, currently, there are 91 communities where 10% or more of livable land area is flooded at least 26 times per year and that number will be likely to nearly double by 2035 (*14*). A primary impact of nuisance flooding in communities is rendering roads impassable or difficult to pass so frequently that routine transport needs are not met. Although the importance of transportation disruptions from nuisance flooding is broadly understood, few studies (*16*) have quantitatively addressed how nuisance flooding impacts roadway flooding.

The goal of this study is to broadly understand the type and extent of roadway infrastructures that are vulnerable to nuisance flooding and the transportation impacts now and in the future due to SLR. This study focuses on the East Coast of the United States and includes all coastal states from Maine to Florida, including the Gulf Coast of Florida and Key West. In this study, we determine the number and length of roadway segments (not including bridges, tunnels, or causeways) at risk from tidal nuisance flooding. The number of days per year that these vulnerable roadways are flooded from nuisance flood is estimated for current, mid-century, and late century. Current traffic counts are combined with the total hours per year that the roads are inundated to determine annual transportation impacts for these same periods. Results are summarized by each state and differentiated by the road type using the Federal Highway Administration (FHWA) functional classifications.

Data Sources

Roadway data for 2015 were obtained from the FHWA. The FHWA requires all states (including the District of Columbia) to annually submit geographical information system (GIS) data on the extent, condition, performance, use, and operating characteristics of all of the nation's public roads, regardless of ownership, to the Highway Performance Monitoring System (HPMS) (*17*). This study uses the HPMS functional class (FC), average annual daily traffic (AADT), facility type (FT), and structure type (ST) data. HPMS functional classes definitions are interstate (FC 1), principal arterial (other freeways and expressways) (FC 2), principal arterial (other) (FC 3), minor arterial (FC 4), major collector (FC 5), minor collector (FC 6) and local roads (FC 7). The facility type (FT) provides the operational characteristic of the roadway and was used to distinguish ramps (FT 4) and individual roads of a multi-road facility that are not used for determining the primary length for the facility (FT 6). FT 6 allows states to more efficiently record interstate information. Structure type (ST) identifies roadway sections that are bridges (ST 1), tunnels (ST 2), or causeways (ST 3). All geospatial data were projected in NAD_1983_2011_Contiguous_USA_Albers, and analysis was done using ArcGIS. States included in this study include Connecticut, Delaware, District of Columbia, Florida, Georgia, Maine, Maryland, Massachusetts, New Hampshire, New Jersey, New York, North Carolina, Pennsylvania, Rhode Island, South Carolina, and Virginia.

Nuisance flooding areas were identified using the National Oceanic and Atmospheric Administration's (NOAA) flood frequency maps that delineate areas based on ground elevation for which there is potential nuisance flooding. Nuisance flooding events correspond to those coastal floods that would cause "minor" impacts to infrastructure as determined locally by the National Weather Service. Most, but not all, of the coastal tide gauges with such elevation thresholds are monitored along with riverine gauges by NOAA (*18*). NOAA used 30-cm LIDAR digital elevation data to develop these maps, which are a spatial interpolation of elevations at or below gauge-specific elevation threshold for nuisance impacts (*19*).

Verified hourly water levels were used in this study and are available from NOAA's Center for Operational Oceanographic Products and Services (*20*). Historic annual flood frequencies are based upon counts of discrete days and total number of hours per year above local minor-flood thresholds (where such thresholds exist) similar to methods of Sweet et al. (*21*). Long-term gauges with hourly data extending prior to 1950 continue to be monitored on an annual basis by NOAA (*21, 22*), and 21 of these tide gauges along the U.S. East Coast and Southern Florida Gulf Coast provided spatial flood frequency information for this study (see Figure 1).

Analysis Methods

Road Segment and AADT Determination

The HPMS data were used to identify roadway segments. Because states submit their own data to FHWA for the HPMS, some aspects of the data in the HPMS may vary slightly from state to state. To account for differences in HPMS data among states, adjustments were made to ensure that data were comparable across the study region. Roadway segments that are bridges, tunnels, or causeways were identified and removed from the analysis because these assets' elevations frequently differed from the ground elevation or the elevations were not available. The definition of a road segment differed by state. For this analysis, road segments

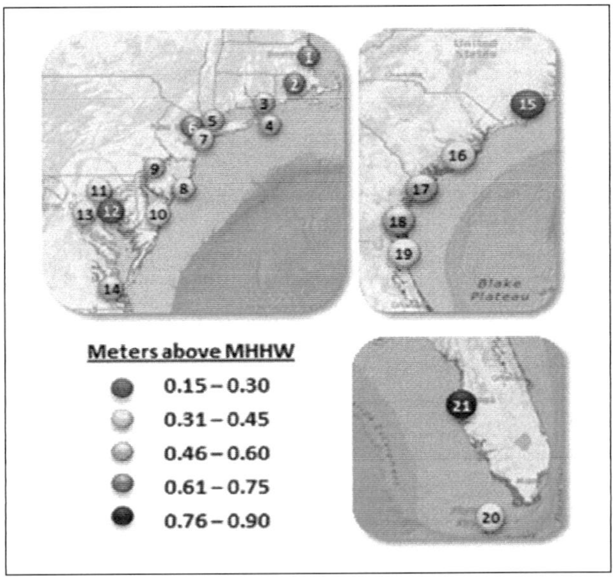

Figure 1. NOAA water level (tide gauge) locations numbered as listed in Table 3 and showing NOAA-defined nuisance-flood levels above the 1983–2001 MHHW (mean higher high water) datum (colored dots) and coastal areas below the nuisance elevations and potentially at risk during nuisance-flood events (red contours) provided by www.csc.noaa.gov/digitalcoast/tools/slrviewer and adapted from Sweet and Park (*8*).

are either defined as the section of road between two ramps for FC 1 and 2 or as the road section between two intersections, not accounting for one-way roads, for FC 3 to 7. FC 1 and 2 segments were defined as a single segment between ramps, created by joining and/or splitting HPMS segments.

The AADT values reported by each state were used to estimate AADT for this study's segments. The AADT values were determined by a distance-weighted average approach. First, vehicle miles traveled (VMT) was determined for each segment (or sub-segment, if split) in the HPMS:

$$VMT = AADT * Segment\_Length_{HPMS} \qquad (1)$$

where $Segment\_Length_{HPMS}$ is the distance of the HPMS-defined segment (or sub-segment).

When multiple segments were joined to create a new segment for this study, the VMT values from the segments were totaled (VMT_{Total}) and a new, length-weighted AADT was determined as

$$AADT_{weighted} = VMT_{Total} / Segment\_Length_{Total} \qquad (2)$$

where $Segment\_Length_{Total}$ is the distance of the segment created for this study.

FC 1 and 2 segments were defined as a single segment between ramps, created by joining and/or splitting HPMS segments. Some states assigned total bidirectional AADT on divided highways to one side of the road and zero to the other side. The zero side was identified using HPMS field facility type 6 (FT6). This was a reporting approach commonly used by Connecticut, Delaware, Florida, Rhode Island, and South Carolina. For all states, first, the non-FT6 roads were split at ramps. VMT was calculated for each segment (Equation 1). Then, segments were joined, creating a single segment between each intersection with a new, length-weighted AADT (Equation 2). FT6 roads were segmented at the same points as the non-FT6 roads by finding the nearest point on the FT6 line to the vertices of the non-FT6 segments. FT6 and non-FT6 segments were paired one-to-one based on the closest FT6 segment to the midpoint of the non-FT6 road, within 175 ft (53.3 m). For non-FT6 roads with a paired segment, half of the AADT was assigned to the FT6 road and half to the pair; segments without a pair remained the same. FT6 segments were then split at ramps, given a VMT factor, and joined between ramps with a new, length-weighted AADT.

In Florida's HPMS submission, ramps were frequently unconnected from highways, sometime separated by hundreds of feet. To account for this, the closest point on the nearest highway segment to the end of each ramp was determined. Highway segments were split at these points, rather than the intersection of the ramps and the road segments. Because South Carolina's 2015 HPMS submission did not include ramps, South Carolina's 2014 ramp submission was used for the South Carolina's FC 1 and 2 analyses.

FC 3, 4, 5, 6, and 7 segments were developed from the intersections of FC 3, 4, 5, 6, and 7 segments. As needed, the HPMS segments in FC 3, 4, and 5 were split at these intersections and VMT was calculated for each segment (Equation 1). Then, segments were joined, creating a single segment between each intersection with a new, length-weighted AADT (Equation 2). Because FC 6 and 7 segments did not have uniform AADT reporting, only the length and number of FC 6 and 7 segments were used in the flood-risk analysis. For the states that did not submit FC 6 and 7 segments to the HPMS, statewide vulnerability summaries did not include this category of roadways.

Vulnerable Road Segments

After creating a consistent set of road segments, these roadway data were overlaid with NOAA's minor tidal flooding extent maps (flood-risk area). Segments that were completely contained in the flood-risk area or intersected the flood-risk area were identified as roadways at risk from tidal nuisance flooding. The number and length of segments intersecting the flood-risk area were summarized by FC for each state. For FC 1, 2, 3, 4, and 5 segments, the AADT values in the flood-risk area were calculated. Because some states reported zero for AADT on ramps, AADT on ramps in the flood-risk area was not included in the delay analysis.

Tide Gauge High-Water Exceedance Projections

Projections of minor flooding were estimated using a combination of contemporary high-water exceedance probabilities and local sea-level-rise scenarios following the methods of Sweet and Park (*8*). Empirical distributions of observed (tide plus storm surge, sea-level anomalies, etc.) highest daily water levels over 1991–2009 (spanning a 19-year tidal epoch) were detrended to establish a mean sea-level datum epoch centered on year 2000. Local relative sea-level-rise projections of Sweet et al. (*6*), which initiate in year 2000, were used to force time-dependent shifts in the high-water distributions relative to the minor elevation thresholds (shown in Table 3). We assumed stationarity within the high-water distribution, which is typical when projecting flood probabilities (*23, 24*), since the focus is on exceedances generally within the bulk of the distribution (< 1 year recurrence interval with very narrow uncertainty bounds) (*21*) largely dominated by repetitive annual tidal forcing and weather patterns. This analysis yielded the annual number of days of flooding and the annual total hours of flooding for each of the 21 tidal gauges. The number of historic nuisance-flood days was summarized using the 10-year average for 2000 (i.e., 1996–2005) and 2010 (i.e., 2006–2015). For the projected nuisance-flood days and annual hours of flooding, average values were determined for 2010, 2020, 2060, and 2100 based upon the local intermediate low, intermediate and extreme sea-level-rise scenarios of Sweet et al. (*6*). All counts are annual totals estimated as the average decadal (10-year) value centered on year (i.e., 2060 is 2056–2065), except 2100, which is a 5-year average (2096–2100).

Roadway Flood Risk

For each vulnerable roadway segment with a FC of 1 through 5, excluding ramps, the distance to the nearest tide gauge was determined using ArcGIS. Because ramps and FC 6 and 7 had nonuniform reporting of segments and AADT, these were excluded from the analysis. For each roadway segment, the annual vehicle-hours of delay (T_Delay) from nuisance flooding was determined using the flooding duration for the nearest tide gauge for each sea-level-rise scenario (Intermediate Low, Intermediate, and Extreme) and each modeled year (2010, 2020, 2060, and 2100):

$$T\_Delay_{\text{scenario, yr}} = T\_Flood_{\text{scenario, yr}} * AADT / (24\text{ h} / \text{day}) \quad (3)$$

where:

$T\_Delay_{\text{scenario, yr}}$ = the annual vehicle-hours of delay for each scenario and modeled year combination; and

$T\_Flood_{\text{scenario, yr}}$ = the cumulative duration of flooding on an annual basis for each scenario and modeled year combination (h).

The total annual vehicle-hours of delay for each state were determined by summing the delays for each segment in that state.

Results

Roads at Risk from Nuisance Flooding

Tidal nuisance flooding across the East Coast of the U.S. threatens 7,508 miles (12,083 km) of roadways and nearly 15,000 individual roadway segments (Tables 1 and 2). Interstates are impacted throughout the study region. In total, over 400 Interstate roadway (FC 1 and 2) segments are at risk, with an average length of 1 mile (1.61 km). In addition, over 300 FC 1 and 2 access ramps are at risk during nuisance flooding. However, most of the roadways at risk are not interstates. Minor or local roads (FC 6 and 7) comprise most of the impacted segments (65%) and half of the total miles. Over 3000 miles (4828 km) of FC 3, 4, and 5 roadways, which play a critical role in local and regional connectivity, are at risk.

Florida and New Jersey have the largest percentage of roads at risk (4.7% and 4.6%, respectively). The four southernmost states (NC, SC, GA, FL) account for over 60% of the impacted miles and roadway segments. In most cases, Florida roads account for about 25% of the total of vulnerable roadways [1823.1 miles (2934 km); 525 segments]. North Carolina has more impacted roadways [2118.2 miles (3409 km); 5636 segments] than Florida. However, North Carolina has significantly more minor and local roads at risk compared with Florida; Florida has more major roads (FC 1–5) at risk. North Carolina has only 3 miles (4.8 km) of interstates roads (FC 1–2) at risk because their highways are typically located away from the coastline. Among Mid-Atlantic states (MD, DC, DE, NJ, NY, PA, VA), New Jersey has a high number and length of affected interstates. The District of Columbia [17.7 miles (28.5 km); 52 segments] has the fewest roads. The New England states (CT, MA, ME, NH, RI) have a relatively low exposure with the most impacts found in FC 3, 4, and 5 roadways. Of the five coastal New England states, Connecticut has the greatest vulnerability.

The total length and number of segments of roadway at risk varies by region and state. Nine of the 16 study states have more than 1% of their reported road network at risk. Smaller states such as RI and DE have relatively few impacted roads, but their total percentage of roads at risk is quite high at 3.1% and 2.8%, respectively. Shorter coastlines and tidal rivers exposure in GA, PA, NY, MA, and NH relative to the state's entire road network keeps the overall total percentage of roads at risk below 0.6%.

Recent Observations and Future Scenarios of Nuisance Flooding Frequency

Annual frequencies of days with a minor (nuisance) tidal flooding have increased rapidly over the last several decades.

Table 1. Miles of Roadway by State and FC Located in Nuisance-Flood Zones

State	Functional Class 1 & 2	Functional Class 1 & 2 ramps	Functional Class 3, 4, & 5	Functional Class 6 & 7	Total	Percent of total road miles
ME	8.7	0.8	188.6	N/A[a]	198.1	2.8%
NH	0.4	0.3	12.4	21	34.2	0.2%
MA	0	7.6	73	159	239.6	0.5%
RI	1.2	1.3	57.4	5.7	65.6	3.1%
CT	5.9	4	117.1	13.8	140.9	1.7%
NY	32	19.3	133	N/A[a]	184.3	0.6%
NJ	77.6	34.3	416.3	7.6	535.8	4.6%
PA	21	14.9	74.1	N/A[a]	110	0.2%
DE	7.7	0.7	58.5	121.5	188.5	2.8%
MD	5.8	1.7	72	548	627.6	1.7%
DC	5.3	3.3	1.2	8	17.7	1.5%
VA	18.7	11.7	145.9	234	410.3	0.6%
NC	3.2	3.3	315.2	1796.5	2118.2	1.4%
SC	31.6	6.4	262.9	288.6	627.2	0.9%
GA	65.4	4.8	53	102	225.3	0.2%
FL	121	52.9	1195.3	453.9	1823.1	4.7%
Total	405.6	167.3	3175.9	3759.6	7508.4	1.1%

Note: 1 mile = 1.61 km.
[a]FC 6 and 7 roadway data were missing.

Table 2. Number of Roadway Segments by State and Functional Class (FC) Located in Nuisance-Flood Zones

State	Functional Class 1 & 2	Functional Class 1 & 2 ramps	Functional Class 3, 4, & 5	Functional Class 6 & 7	Total
ME	10	3	147	N/A[a]	160
NH	1	1	56	80	138
MA	0	12	352	1139	1503
RI	4	5	89	6	104
CT	16	16	201	28	261
NY	41	43	212	N/A[a]	296
NJ	67	49	567	27	710
PA	25	24	111	N/A[a]	160
DE	11	2	107	347	467
MD	8	4	173	1218	1403
DC	7	14	6	25	52
VA	33	33	199	374	639
NC	3	9	592	5032	5636
SC	21	15	306	531	849
GA	23	13	114	364	514
FL	121	100	1198	525	1944
Total	391	343	4430	9696	14860

[a]FC 6 and 7 roadway data were missing.

In fact, flood frequencies are accelerating in dozens of U.S. East and Gulf Coast communities even as local mean sea-level trends are not necessarily accelerating (*21*). Acceleration in tidal flooding occurs as rising sea levels evolve the nonlinear portion of a water level distribution against a fixed elevation (i.e., nuisance-flood elevation threshold shown in Figure 1 and listed in Table 3), irrespective of whether the SLR is linearly increasing or nonlinearly accelerating (*8*). Comparing decadal-averaged flood frequencies in 2000 (i.e., 1996–2005) to those in 2010 (Table 3), the change is substantial, with an average increase across all sites of about 90%, ranging from about a 25% increase (New London, CT) to upwards of a 350% increase (Key West, FL).

When a nuisance flood occurs, the flooding typically lasts for only part of the day, often for an hour or so near the peak of one or both of the approximately twice-daily (semi-diurnal)

Table 3. Observations and Projections of Annual Nuisance-Flood Days for Various SLR Scenarios at Various Locations[a]

St.	Location Name	Station ID	Nuisance Level	Observed	Obs.	Intermediate Low			Intermediate			Extreme		
			(m, MHHW)	2000	2010	2020	2060	2100	2020	2060	2100	2020	2060	2100
1	Boston, MA	8443970	0.68	2	6	10	53	115	16	155	359	40	364	365
2	Providence, RI	8454000	0.66	1	2	4	36	109	7	155	362	22	365	365
3	New London, CT	8461490	0.60	1	2	4	38	165	6	243	365	21	365	365
4	Montauk, NY	8510560	0.60	1	2	4	50	223	8	262	365	25	365	365
5	Kings Point, NY	8516945	0.52	8	15	23	106	214	37	254	363	75	365	365
6	Battery (NYC), NY	8518750	0.65	2	4	6	50	162	10	197	363	28	365	365
7	Sandy Hook, NJ	8531680	0.45	14	22	43	194	320	66	330	365	121	365	365
8	Atlantic City, NJ	8534720	0.43	13	22	49	214	330	69	334	365	127	365	365
9	Philadelphia, PA	8545240	0.49	6	11	20	137	283	32	305	364	85	365	365
10	Lewes, DE	8557380	0.41	12	21	40	192	323	57	330	365	123	365	365
11	Baltimore, MD	8574680	0.41	7	14	31	202	321	49	330	365	130	365	365
12	Annapolis, MD	8575512	0.29	24	41	87	292	351	122	354	365	224	365	365
13	Washington DC	8594900	0.31	21	30	72	259	336	102	342	365	196	365	365
14	Norfolk, VA	8638610	0.53	5	8	16	142	323	22	318	365	59	365	365
15	Wilmington, NC	8658120	0.25	23	43	71	275	356	108	361	365	204	365	365
16	Charleston, SC	8665530	0.38	16	25	53	183	309	70	311	365	125	365	365
17	Savannah, GA	8670870	0.46	10	17	37	139	265	51	267	365	94	365	365
18	Fernandina Beach, FL	8720030	0.59	1	2	7	48	139	11	151	362	32	364	365
19	Mayport, FL	8720218	0.44	2	6	14	94	245	23	254	365	63	365	365
20	Key West, FL	8724580	0.33	1	4	18	195	354	35	357	365	100	365	365
21	St Petersburg, FL	8726520	0.84	0	0	0	1	15	0	16	355	1	362	365

Note: 1 m = 0.305 ft.
[a]All counts are annual totals estimated as the decadal average decadal (10-yr) value centered on year (i.e., 2000 is 1996–2005), except 2100, which is a 5-year average (2096–2100).

high tides typical along the East and Gulf Coasts. The total hours per year at or above the nuisance-flood threshold at a specific location are closely related to the days per year (via a multiplication factors of 2.5 hours/day on average) (*8*) and reveal similar trends (not shown). Comparing all locations, a general pattern exists ($R^2 \sim 0.7$) as recently explained by Sweet et al. (*13*): where elevation thresholds are lower (i.e., Figure 1), more flood days occurs (e.g., Wilmington, NC); where thresholds are higher such as in St. Petersburg (reflecting hurricane-flood mitigation), nuisance flooding is mostly nonexistent.

With continued local relative sea level rise, the change in nuisance-flood frequency (and depth and extent) is projected to grow most rapidly at all locations, as they have in the past, most notably where local SLR rates are higher (*7, 8*). For example, Figure 2 shows the increases in observed nuisance floods as well as the projected nonlinear increases in nuisance-flood frequency values under a range of scenarios from Sweet et al. (*6*) for Charleston, SC. For the 21 tide gauges in the Eastern U.S., Table 3 provides indicators of the projected increases in flooding for the vulnerable roadways by region. Under the Intermediate scenario, the number of days per year with nuisance flooding will approach saturation (365 days per year with a flood) on average in the decade of 2060 (2056–2065) at Sandy Hook, NJ, (330 days), Atlantic City, NJ, (334 days), Annapolis, MD, (354 days), Washington, DC, (342 days), Wilmington, NC, (361 days), and Key West, FL (357 days).

The elevation threshold and amount of local SLR by 2060 explain 98% of the variability in the days/year metric between locations shown in Table 3 ($R^2 = 0.98$, not shown). As would be expected, the annual flood frequencies (days/year) reaches saturation sooner under the higher sea-level-rise scenarios (e.g., extreme scenario). However, it should be noted that the current trajectory of local mean sea level (and derived flood frequency metric) at most locations are currently tracking the more probable scenarios [e.g., intermediate low scenario of Sweet et al. (*6*)].

Impacts of Nuisance Flooding to Roadway Performance

The total annual vehicle-hours of delay for each state were mapped for each year and scenario (Figure 3). Modeled data for 2010 provide a baseline for comparison. Under baseline conditions, nuisance flooding is already causing delays in all East Coast states with the Southern states having more induced vehicle-hours of delay than the northern states. Total delays currently exceed 100 million hours annually. In the

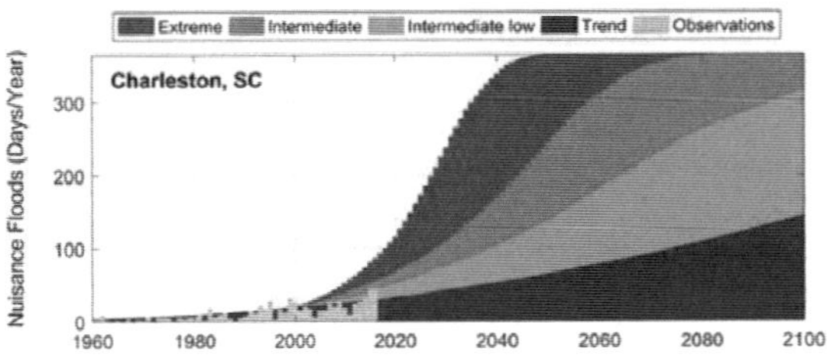

Figure 2. NOAA-defined nuisance-flood levels for Charleston, SC (Station ID 8665530), including observations through 2016 and projections to 2100 based upon the local sea-level trend and the intermediate low, intermediate, and extreme sea-level-rise scenarios of Sweet et al. (6).

future, nuisance-flooding-induced vehicle-hours of delay, or congestion, will increase for all states' coastal roads as compared with baselines conditions. Congestion also increases as scenarios become more extreme. The intermediate low scenario results in nearly 160 million vehicle-hours of congestion across the East Coast by 2020 (an 85% increase from 2010); 1.2 billion vehicle-hours by 2060 (126% increase from 2010); and 3.4 billion vehicle-hours by 2100 (392% increase from 2010) (Figure 2). Under the intermediate and extreme scenarios, vehicle-hours of delay increase 124% and 236% from 2010 to 2020, respectively. Schrank et al. (*25*) estimated that in 2014, total travel delay for the U.S. was 6.9 billion hours. Vehicle-hours of delay from nuisance flooding on the East Coast alone will exceed that level by 2100 for the intermediate scenario and by 2060 for the extreme scenario.

Nuisance-flooding-induced congestion varies widely by region and state. Flooding increases in the Southern states will occur faster that for the Mid-Atlantic or New England states. Together, the four Southern states make up 36% of nuisance-flooding-induced congestion in 2020, rising to 51% in 2100 under the intermediate low scenario. By contrast, the seven Mid-Atlantic states make up 59% in 2020, dropping to 45% in 2100; the five New England states make up 5% in 2020, dropping to 4% in 2100. Virginia is the only state to reach 25 million vehicle-hours of delay by 2020 in the intermediate low scenario. However, for the more extreme scenarios in mid- to late-century, Florida will consistently experience the greatest amount of congestion of any state examined in this study. In the most conservative (intermediate low) scenario, Florida could see an increase of 1 billion hours of delay by 2100 compared with 2010.

Discussion

FHWA now requires state DOTs and metropolitan planning organizations (MPOs) to calculate several metrics related to congestion and reliability of the national highway system (i.e., level of travel-time reliability, annual hours of peak-hour excessive delay per capita, truck travel-time reliability), and to set targets for these metrics and achieve progress toward targets (*26*). Nuisance flooding has the potential to significantly impact coastal states' abilities to meet their targets, if set without consideration of future changes. Because state DOTs and MPOs are not required to establish targets for the first performance period (which ends in 2021) until 2018, considering the impact of nuisance flooding—particularly at the 2020 levels predicted in this study—in target-setting may help state DOTs and MPOs set more achievable and realistic targets.

FHWA also requires state DOTs to follow a process for maintaining physical assets in a state of good repair over their lifecycle (*27*). As part of determining the lifecycle of an asset, FHWA says that state DOTs should consider future climate-related changes. Nuisance flooding may cause minor damage to infrastructure that can rapidly lead to the deterioration of assets, especially when damage is increasingly recurrent, as predicted by this study (*28*).

This study uses a regional analysis approach to consistently identify roadways vulnerable to nuisance flooding and its impacts on transportation delays. Datasets that provide more detailed spatial information or a finer temporal resolution may improve the understanding of flooding impacts to specific road segments. This study does not include existing, planned, or potential protective structures or adaptation strategies to reduce roadway flooding. These measures could reduce the frequency and duration of flooding. All FC 1–5 (excluding ramps) road segments were considered in the roadway performance analysis; however, the distance from the nearest tide gauge varied. A total of 44% of segments were less than 25 miles (40 km) from the nearest tide gauge and 65% of segments were less than 50 miles (80 km) from the nearest tide gauge, but segments ranged from 0.02 to 343 miles (0.03 to 552 km) away. As scenario-based modeling that uses a greater number of tide gauges becomes available, future studies will be able to more accurately predict flood timing on specific road segments.

The time at which roads are impassable or difficult to pass determines the extent to which routine transport needs are impeded. On the East Coast, the probability of high tide occurring at a particular hour is about the same over the course of a year due to the progressive nature of the semidiurnal tide (i.e., 12.4 oscillation). Thus, AADT values are adequate to resolve this study's annual hours of delay metric. However, AADT is not uniformly distributed throughout the day and peak hours [e.g., morning (6–10 am) and evening commute (3–8 pm) (*26*)] are likely to experience more traffic than nonpeak hours. Furthermore, the timing and duration of roadway inundation will vary within the minor-flood area and differentially impact segments.

Summary

SLR and flooding due to storm surge are understood to make coastal roadways more vulnerable and less functional.

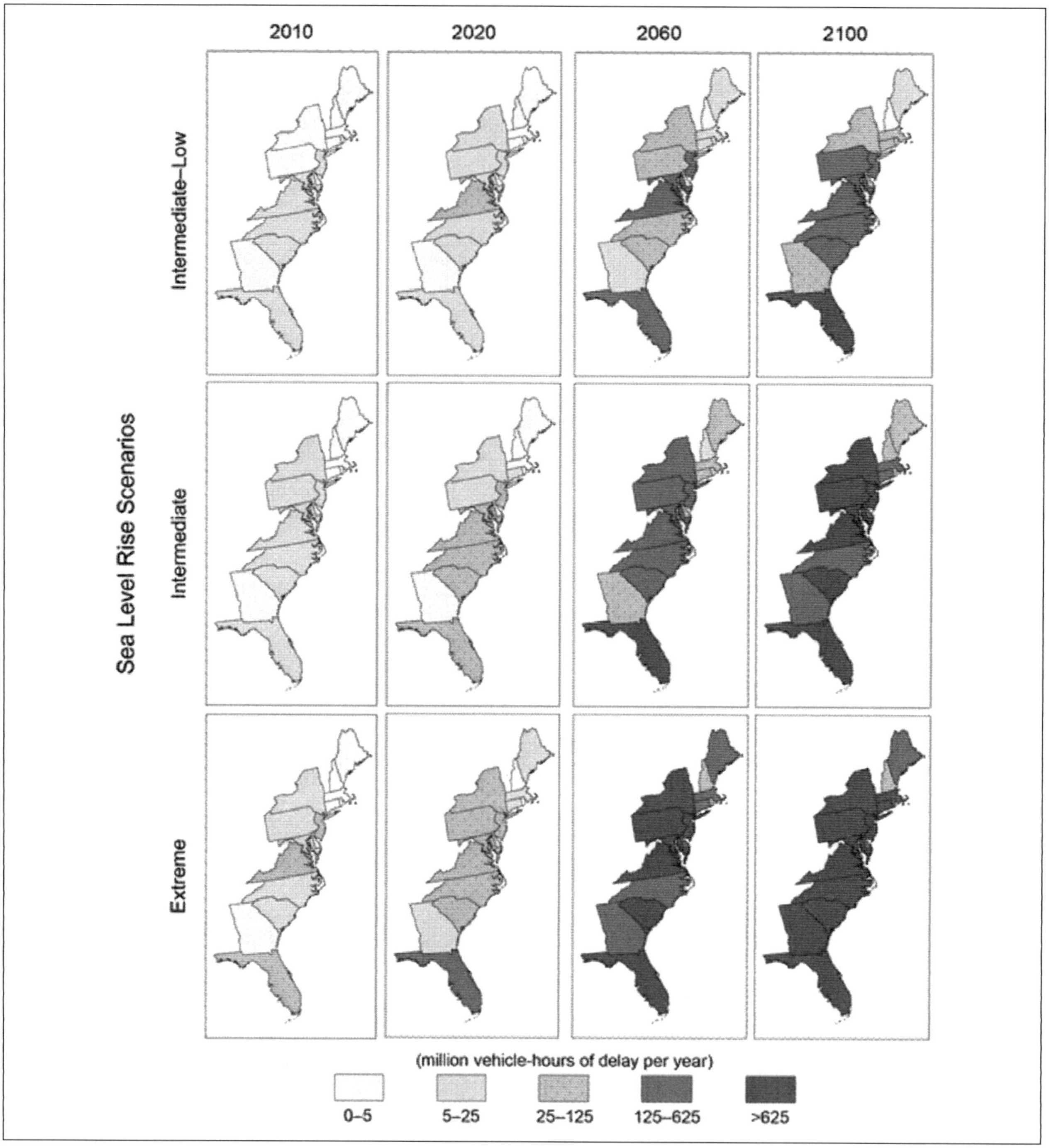

Figure 3. Annual vehicle-hours of delay for FC 1–5 roads due to tidal nuisance flooding by state, year (2010, 2020, 2060, 2100) using decadal average decadal (10-year) values except 2100, which is a 5-year average (2096–2100) and intermediate low, intermediate, and extreme sea-level-rise scenarios of Sweet et al. (6).

This study identifies nuisance flooding as an important, additional stressor that should be considered when characterizing transportation infrastructure's vulnerability to SLR and identifying resiliency measures. This study documents the number of vehicles and hours affected on flooded roads in the NOAA-designated minor-flood area. The results indicate that tidal nuisance flooding across the East Coast threatens thousands of miles of roadways including hundreds of miles of interstate roadways. Over the past 20 years, the frequency of nuisance floods has nearly doubled and is projected to continue to increase at all locations. The total induced vehicle-hours of delay due to nuisance flooding currently exceed 100 million hours annually and could exceed a billion vehicle-hours by 2060. In addition to the reported delays, flooding affects transportation system performance with direct and indirect impacts to logistics, safety, local economies, and emissions.

Future analyses are needed to more broadly understand transportation impacts from tidal nuisance flooding and to better understand impacts in specific regions. Roads outside of NOAA's minor-flood area may also be increasingly vulnerable to flooding. Furthermore, traffic on roads surrounding minor-flood areas may also be delayed as these road accommodate traffic that would otherwise flow on flooded roads. In some regions, there are existing, planned, or potential protective structures or adaptation strategies to reduce roadway flooding that were not considered in this study. These measures could reduce the frequency and duration of flooding and improve a community's ability withstand or recover from tidal nuisance flooding.

Acknowledgments

This work was conducted through the Infrastructure and Climate Network (ICNet) project funded by the U.S. National Science Foundation research grant (CBET-1231326). The authors also acknowledge the help of the University of New Hampshire Center for Infrastructure Resilience to Climate (UCIRC).

References

1. Peterson, T. C., M. McGuirk, T. G. Houston, A. H. Horvitz, and M. F. Wehner. *Special Report 290: The Potential Impacts of Climate Change on U.S. Transportation:* Climate Variability and Change with Implications for Transportation. Transportation Research Board of the National Academies, Washington, D.C., 2008.
2. Douglass, S. L., B. M. Webb, and R. Kilgore. 2014. *Highways in the Coastal Environment: Assessing Extreme Events*. Report FHWA-NHI-14–006. FHWA, U.S. Department of Transportation, 2014.
3. Blase, A. 2016. Projected Flood Impacts from Sea Level Rise, Tides, and Spatially Variable Storm Surge on Roadways in Norfolk and Virginia Beach, VA.
4. Poulter, B., and P. N. Halpin. Raster Modelling of Coastal Flooding from Sea-Level Rise. *International Journal of Geographical Information Science*, Vol. 22, No. 2, 2008, pp. 167–182. doi:10.1080/13658810701371858.
5. Belfield, S. S., and B. McFarlane. *Sea Level Rise and Storm Surge Impacts to Roadways in Hampton Roads*. Report T16-03. Hampton Roads Transportation Planning Organization, Chesapeake, VA, May 2016.
6. Sweet, W. V., R. E. Kopp, C. Weaver, J. Obeysekera, R. Horton, E. R. Thieler, and C. Zervas. *Global and Regional Sea Level Rise Scenarios for the United States*. NOAA Tech. Rep. NOS CO-OPS 083. National Oceanic and Atmospheric Administration, 2017. tidesandcurrents.noaa.gov/publications/techrpt83_Global_and_Regional_SLR_Scenarios_for_the_US_final.pdf. Accessed Jul. 30, 2017.
7. Ezer, T., and L. P. Atkinson. Accelerated Flooding along the U.S. East Coast: On the Impact of Sea-Level Rise, Tides, Storms, the Gulf Stream, and the North Atlantic Oscillations. *Earth's Future*, Vol. 2, 2014, pp. 362–382. doi:10.1002/2014EF000252.
8. Sweet, W. V., and J. Park. From the Extreme and the Mean: Acceleration and Tipping Point of Coastal Inundation from Sea Level Rise. *Earth Futures*, Vol. 2, 2014, pp. 579–600. doi:10.1002/2014EF000272.
9. Moftakhari, H. R., A. AghaKouchak, B. F. Sanders, and R. A. Matthew. Cumulative Hazard: The Case of Nuisance Flooding. *Earth's Future*, Vol. 5, 2017, pp. 214–223. doi:10.1002/2016EF000494.
10. Moftakhari, H. R., A. AghaKouchak, B. F. Sanders, D. L. Feldman, W. Sweet, R. A. Matthew, and A. Luke. Increased Nuisance Flooding along the Coasts of the United States Due to Sea Level Rise: Past and Future. *Geophysical Research Letters*, Vol. 42, 2015, pp. 9846–9852. doi:10.1002/2015GL066072.
11. Ray, R. D., and G. Foster. Future Nuisance Flooding at Boston Caused by Astronomical Tides Alone. *Earth's Future*, Vol. 4, No. 12, 2016, pp. 578–587. doi: 10.1002/2016EF000423.
12. Vandenberg-Rodes, A., H. R. Moftakhari, A. AghaKouchak, B. Shahbaba, B. F. Sanders, and R. A. Matthew. Projecting Nuisance Flooding in a Warming Climate Using Generalized Linear Models and Gaussian Processes. *Journal of Geophysical Research: Oceans*, Vol. 121, No. 11, 2016, pp. 8008–8020. doi: 10.1002/2016JC012084.
13. Sweet, W. V., J. J. Marra, and G. Dusek. 2016 State of U.S. High Tide Flooding and a 2017 Outlook. Supplement to State of the Climate: National Overview for May 2017, published online June 2017. www.ncdc.noaa.gov/monitoring-content/sotc/national/2017/may/2016_StateofHighTideFlooding.pdf. Accessed July 30, 2017.
14. Dahl, K. A., E. Spanger-Siegfried, A. Caldas, and S. Udvardy. Effective Inundation of Continental United States Communities with 21st Century Sea Level Rise. *Elementa: Science of the Anthropocene*, Vol. 5, 2017, 37. doi:10.1525/elementa.234.
15. Kulp, S., and B. H. Strauss. Rapid Escalation of Coastal Flood Exposure in US Municipalities from Sea Level Rise. *Climatic Change*, Vol. 142, No. 3–4, 2017, pp. 477–489. doi:10.1007/s10584–017–1963–7.
16. Buchanan, M. K., M. Oppenheimer, and R. E. Kopp. Amplification of Flood Frequencies with Local Sea Level Rise and Emerging Flood Regimes. *Environmental Research Letters*, Vol. 12, No. 6, 2017, 111001.
17. Federal Highway Administration. *Highway Performance Monitoring System Field Manual*. Office of Highway Policy Information, 2016. www.fhwa.dot.gov/policyinformation/hpms/fieldmanual/hpms_field_manual_dec2016.pdf. Accessed Jul. 30, 2017.
18. National Weather Service. *National Observations: Water*. National Oceanic and Atmospheric Administration. http://water.weather.gov/ahps. Accessed July 30, 2017.
19. National Ocean Service, Office of Coastal Management. *Coastal Flood Exposure Mapper*. National Oceanic and

Atmospheric Administration. www.coast.noaa.gov/floodexposure/#/select Accessed July 30, 2017.

20. National Ocean Service, Center for Operational Oceanographic Products and Services. *Tides and Currents Map*. National Oceanic and Atmospheric Administration. https://tidesandcurrents.noaa.gov/ Accessed July 30, 2017.
21. Sweet, W. V., J. Park, J. J. Marra, C. Zervas, and S. Gill. Sea level rise and nuisance flood frequency changes around the United States, NOAA Tech. Rep. NOS CO-OPS 73, 53 pp. https://tidesandcurrents.noaa.gov/publications/NOAA_Technical_Report_NOS_COOPS_073.pdf. 2014. Accessed July 30, 2017.
22. Sweet, W. V., and J. J. Marra. 2015 State of U.S. Nuisance Tidal Flooding. Supplement to *State of the Climate: National Overview* for May 2016, published online June 2016. www.ncdc.noaa.gov/monitoring-content/sotc/national/2016/may/sweet-marra-nuisance-flooding-2015.pdf. 2016. Accessed July 30, 2017.
23. Tebaldi, C., B. H. Strauss, and C. E. Zervas. Modeling Sea Level Rise Impacts on Storm Surges along US Coasts. *Environmental Research Letters*, Vol. 7, 2012, 014032. doi:10.1088/1748–9326/7/1/014032.
24. Hall, J. A., S. Gill, J. Obeysekera, W. Sweet, K. Knuuti, and J. Marburger. *Regional Sea Level Scenarios for Coastal Risk Management: Managing the Uncertainty of Future Sea Level Change and Extreme Water Levels for Department of Defense Coastal Sites Worldwide*. U.S. Department of Defense, Strategic Environmental Research and Development Program, 2016.
25. Schrank, D., B. Eisele, T. Lomax, and J. Bak. *2015 Urban Mobility Scorecard*. Texas A&M Transportation Institute and INRIX, Inc. 2015. https://static.tti.tamu.edu/tti.tamu.edu/documents/mobility-scorecard-2015.pdf. Accessed July 30, 2017.
26. Federal Highway Administration. *National Performance Management Measures; Assessing Performance of the National Highway System, Freight Movement on the Interstate System, and Congestion Mitigation and Air Quality Improvement Program*. Final Rule 82 FR 5970.
27. Federal Highway Administration. *Asset Management Plans and Periodic Evaluations of Facilities Repeatedly Requiring Repair and Reconstruction Due to Emergency Events*. Final Rule 81 FR 73196.
28. ARA, Inc. *Guide for Mechanistic-Empirical Design of New and Rehabilitated Pavement Structures*. National Cooperative Highway Research Program.

The Special Task Force on Climate Change and Energy (A0020T) peer-reviewed this paper (18-02608).

Article

TRR
JOURNAL OF THE TRANSPORTATION RESEARCH BOARD

Transportation Research Record
2018, Vol. 2672(2) 11–22

DOI: 10.1177/0361198118757441
journals.sagepub.com/home/trr
SAGE

Adaptation Planning to Mitigate Coastal-Road Pavement Damage from Groundwater Rise Caused by Sea-Level Rise

Jayne F. Knott[1], Jo Sias Daniel[1], Jennifer M. Jacobs[1], and Paul Kirshen[2]

Abstract

Sea level in coastal New England is projected to rise 3.9–6.6 ft (1.2–2.0 m) by the year 2100. Many climate-change vulnerability and adaptation studies have investigated surface-water flooding from sea-level rise (SLR) on coastal-road infrastructure, but few have focused on rising groundwater. Groundwater modeling in New Hampshire's Seacoast Region has shown that SLR-induced groundwater rise will occur three to four times farther inland than surface-water flooding, potentially impacting 23% of the region's roads. Pavement service-life has been shown to decrease when the unbound layers become saturated. In areas where groundwater is projected to rise with SLR, pavements with groundwater 5.0 ft (1.5 m) deep or less are at risk of premature failure as groundwater moves into the pavement's underlying unbound layers. In this study, groundwater hydrology and multi-layer elastic pavement analysis were used to identify two case-study sites in coastal New Hampshire that are predicted to experience pavement service-life reduction caused by SLR-induced groundwater rise. Various pavement structures were evaluated to determine adaptation feasibility and costs to maintain the designed service-life in the face of rising groundwater. This investigation shows that relatively simple pavement structural modifications to the base and asphalt concrete (AC) layers of a regional corridor can eliminate the 80% to 90% service-life reduction projected with 1.0 ft SLR (year 2030) and will delay pavement inundation by 20 years. Pavements with adequate base-layer materials and thickness require only AC thickness modification to avoid premature pavement failure from SLR-induced groundwater rise.

Sea-level rise (SLR) can cause erosion and storm-surge damage, flood coastal communities, and damage coastal infrastructure (*1*). Many studies have investigated SLR-induced surface-water flooding and storm surge on road infrastructure (*2*, *3*) but few have investigated the effect of SLR-induced groundwater rise on coastal-road pavement systems. As groundwater rises, the unsaturated zone decreases (*4*) and groundwater will impact roads, septic systems, underground utilities, and foundations in areas where the groundwater is currently near the ground surface (*5*).

The effect of rising groundwater on pavement performance has been investigated. Full-scale pavement tests conducted in Florida test pits found a more than 50% reduction in the resilient modulus of granular subgrade layers when the groundwater table rises 24 in. (61 cm) to the bottom of the base layer (*6*). A study of SLR and increased precipitation on roads along the Gold Coast in Australia found that approximately 1.9 miles (3.0 km) (32%) of roadway will be at high risk of failure by 2070 when groundwater is predicted to be less than 3.3 ft (1.0 m) below the pavement surface (*7*). Similarly, SLR-induced groundwater rise was found to reduce coastal-road pavement life by 50% from fatigue-cracking distress, and by up to 90% from rutting distress when groundwater moves into the pavement base layers (*8*).

Climate-change adaptation studies assess the vulnerability of assets, choose appropriate climate-change scenarios, identify adaptation strategies, and recommend actions that may be robust or flexible (*9*). Studies have shown that some pavement structures are more resilient to rising groundwater than

[1]University of New Hampshire, Durham, NH
[2]University of Massachusetts Boston, Integrated Science Complex, Boston, MA

Corresponding Author:
Address correspondence to J. F. Knott: jfk1011@wildcats.unh.edu

others (*10, 11*), but to date, no studies have quantified the cost and benefits of strategies to mitigate pavement impacts from SLR-induced groundwater rise. The implementation of targeted adaptation actions to vulnerable coastal-road infrastructure will save money through damage avoidance (*8*).

The objective of this research was to identify pavement structures that will preserve pavement service-life in the face of SLR-induced groundwater rise. It provides a quantitative assessment of pavement strains for projected groundwater levels and compares service-life reductions in terms of fatigue and rutting across different pavement structures. Pavement construction costs and road-surface elevations were evaluated to determine cost-effective strategies to prevent premature pavement failure as rising groundwater weakens the pavement supporting layers and inundates the pavement surface. Whereas rising temperatures may also weaken pavements (*12*), this investigation considered only the effects of SLR-induced groundwater rise on pavement service-life. Fatigue cracking and rutting were evaluated independently, and it was assumed that no pre-existing distresses existed in the pavement evaluation sites. In reality, these distresses are coupled and exist to some degree in the pavement structure; consequently, pavement failure will likely occur sooner than predicted in this analysis. SLR-induced groundwater-rise projections are limited to the SLR scenarios chosen for this study.

Background

Study Area

The study area was 158 miles^2 (410 km^2) of the New Hampshire (NH) Seacoast Region where there is approximately 18 miles (29 km) of shoreline (*13*). The study area is bounded by the Piscataqua Estuary in the north, Great Bay and the Squamscott River in the west, the Massachusetts border in the south and the Gulf of Maine in the east (Figure 1).

Groundwater Modeling

Groundwater modeling, using USGS MODFLOW-2005 (*14*), was conducted to identify the NH groundwater-rise zone (NH-GWRZ) within the study area (Figure 1). This is the land area where groundwater is projected to rise with SLR (*8*). SLR scenarios of 1.0 ft (0.3 m), 2.7 ft (0.8 m), 5.2 ft (1.6 m) and 6.6 ft (2.0 m) in early-century (2030), mid-century (2060) and end of century (2090 and 2100) were simulated (*15*). Results demonstrate that groundwater rise will extend more than three to four times farther inland than surface-water impacts (*8*). Approximately 23% (146 miles, 235 km) of the region's roads, or 30% of NH-GWRZ roads, currently have groundwater less than 5.0 ft (1.5 m) below the road surface and are vulnerable to pavement damage. In addition, road-surface inundation is projected to occur in approximately 14% (20 miles, 32 km) of these roads with 6.6 ft (2.0 m) SLR.

Case-Study Sites

In a previous study, the pavement service-life reduction was determined at five pavement evaluation sites using simulated groundwater levels corresponding with the SLR scenarios (*8*). All pavements are projected to experience service-life reductions with groundwater rise, the magnitude of which depends on the location, local hydrogeology, and pavement structure (*8*). Two case-study sites, Route 286 and Gosling Road, were chosen for this study based on projected service-life reduction and criticality with input from the New Hampshire Department of Transportation (NHDOT), regional planning commissioners, and the New Hampshire Department of Environmental Services (NHDES) (Figure 1).

Route 286 is an east–west 2-lane highway spanning Seabrook, NH and Salisbury, Massachusetts. It is an evacuation route from a popular tourist area along the Atlantic shoreline and the evaluation site is located approximately 1.3 mi (2.1 km) from the coast and within 400 ft (122 m) of a coastal wetland. Gosling Road, located in Newington, NH, is a four-lane local road providing access to Portsmouth International Airport at Pease and a local shopping mall. The evaluation site is located 0.8 mi (1.3 km) from the Piscataqua Estuary.

Materials, Methods, and Data

Pavement Evaluation

Pavement Profiles and Material Properties. Pavement profiles were constructed with information from NHDOT construction plans (*16*) and boring logs from the New Hampshire Geological Survey and NHDES (*8*). The material properties for the existing pavement structure with the current groundwater table are presented in Table 1. The AC, base and subbase material resilient modulus values were estimated from AASHTO correlation charts (*17*) using NHDOT layer coefficients determined from field tests (*18, 19*). Subgrade resilient modulus values were obtained from laboratory tests of subgrade soil types typically found in NH (*20*). Poisson's ratio of unbound materials was assigned to each lithographic layer per its AASHTO soil classification (*17*). The soil beneath the groundwater table is fully saturated and the unsaturated soil above the groundwater table is assumed to be at optimum moisture content used for pavement design. The ratio of resilient modulus at full saturation to resilient modulus at optimum moisture content (M_{Rsat}/M_{Ropt}) is assumed to be 0.5 for the unbound materials based on MEPDG guidance (*21*). The AC dynamic modulus fluctuates with temperature, but was assumed to be constant to isolate the effect of rising groundwater.

Multi-Layer Elastic Analysis. KENLAYER computer software (*22*) was used to determine changes in pavement response to loading resulting from SLR-induced groundwater rise under

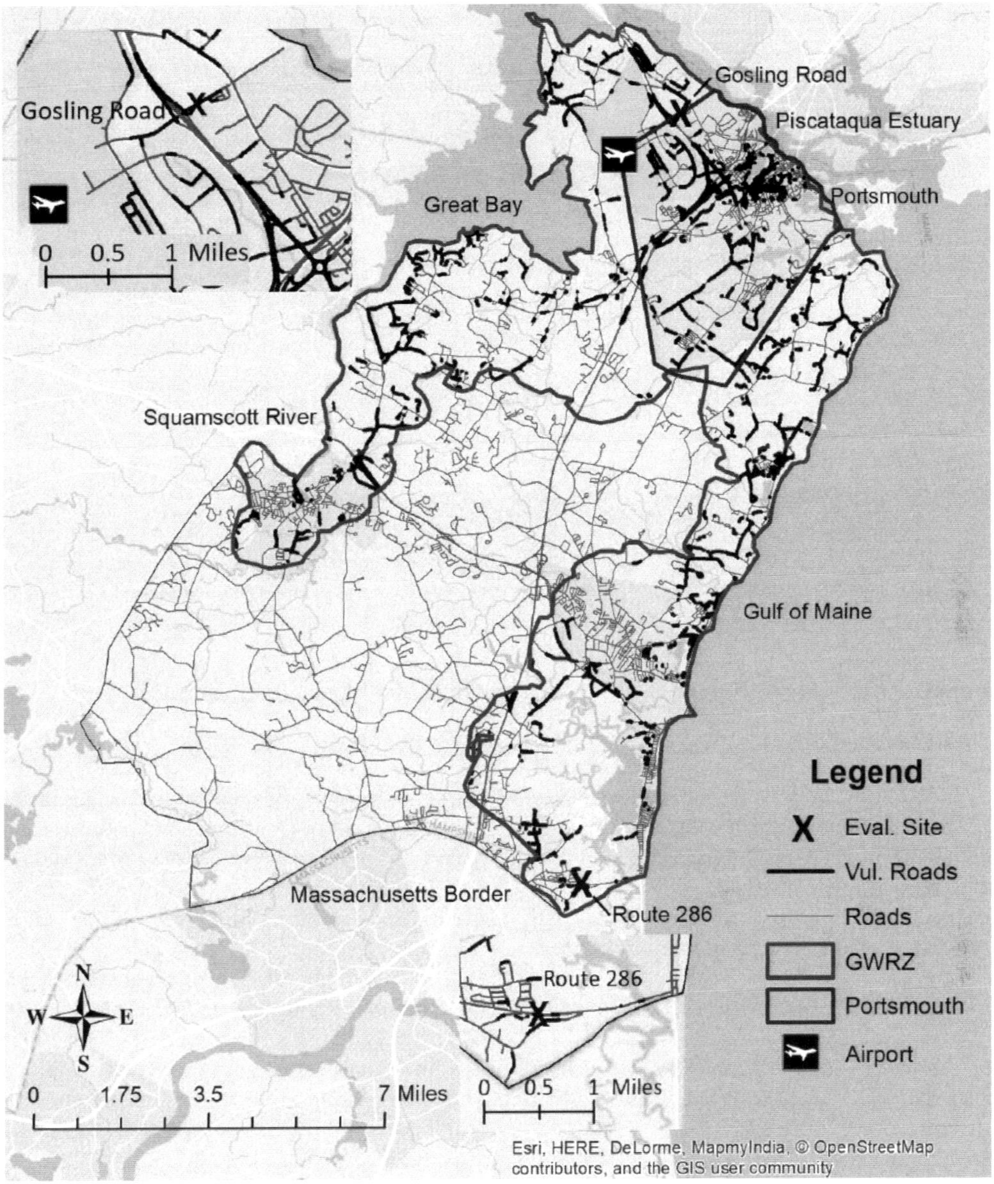

Figure 1. Study area in coastal NH. The NH-GWRZ (Groundwater-Rise Zone) is the projected area of SLR-induced groundwater rise. Two case-study sites, Gosling Road and Route 286, were chosen for the evaluation of adaptation strategies. Vulnerable roads are shown in **BOLD**.

full bond and full slip interlayer conditions. KENLAYER is based on multi-layer elastic theory that assumes each pavement layer is continuous, homogeneous, isotropic and linearly elastic. Loading was kept constant at 9,000 lb (4,082 kg) single axle single tire with a contact radius of 4.89 in. (12.4 cm) using the methodology presented in Knott et al. (*8*). The number of cycles to failure (N_f) for fatigue cracking and rutting were determined from the horizontal tensile strain calculated at the bottom of the AC layer and the vertical compressive strain calculated at the top of the subgrade using transfer functions developed by the Minnesota Department of Transportation (*23*). Pavement service-life reduction is represented by the ratio of the N_f for the predicted groundwater condition (N_{fp}) to the N_f for the current groundwater condition (N_{fc}). After analyzing the existing structure, structures were modified and fatigue and rutting N_f s were calculated with the projected groundwater levels to determine viable adaptation strategies.

Table 1. Material Properties and Input Parameters for Multi-Layer Elastic Analysis at Route 286 and Gosling Road with Saturation

Road and tier classification	Layer type	Material	AASHTO class	Resilient modulus (psi)	Poisson's ratio	Thickness (in.)	U/S
Route 286 (regional corridor)	Asphalt	HMA	na	325,000	0.35	5.5	U
	Base	Gravel	A-1-a	15,000	0.35	16	U
	Subgrade	Sand	A-3	3,800	0.30	1	U
				1,900	0.30	105	S
		Sand and gravel	A-1-b	19,250	0.35	294	S
		Clay	A-7-6	1,500	0.45	180	S
		Till	A-2-7	3,250	0.25	12	S
		Bedrock	na	725,000	0.20	infinite	S
Gosling Road (local road)	Asphalt	HMA	na	325,000	0.35	4	U
	Base 1	Crushed gravel	A-1-a	21,000	0.35	12	U
	Base 2	Gravel	A-1-a	15,000	0.35	12	U
	Subbase	Sand	A-3	4,500	0.30	8	S
	Subgrade	Silt	A-4	8,250	0.33	40	S
		Gravel	A-1-a	19,250	0.35	24	S
		Sand and silt	A-2-4	4,500	0.25	24	S
		Till	A-2-7	10,250	0.25	infinite	S

Note: psi = pounds per square inch; kPa = kilopascal; 1 in. = 25.4 mm; 1 psi = 6.89 kPa; na = not applicable; U = unsaturated; S = saturated; HMA = hot mix asphalt.

Adaptation Options

Adaptation options chosen for analysis were based on input from NHDOT and local planners. These included changes in layer thicknesses and base/subbase material types with the goal to meet or exceed a performance threshold. Increases in road-surface elevation with the new structure were considered beneficial because they delay road-surface inundation from surface water or groundwater. Route 286 crosses a coastal wetland and does not have many adjacent properties; consequently, it is possible to increase the road-surface elevation without major adjacent-road modifications. Gosling Road has adjacent properties, including a shopping mall, and a nearby overpass making increases in road-surface elevation impractical. Consequently, adaptation options that raised the road surface at the Gosling Road site were not considered in this analysis.

The adaptation options analyzed for Route 286 and Gosling Road are presented in Table 2. These options are not exhaustive, but are used to illustrate the methodology.

Pavement structures that meet the following performance threshold are considered viable adaptation strategies:

$$\frac{N_{fn}}{N_{fe}} \geq 1 \quad (1)$$

where N_{fn} / N_{fe} is the relative pavement life, N_{fn} is the number of cycles to failure with the new pavement structure and projected groundwater levels, and N_{fe} is the number of cycles to failure with the existing pavement structure and existing groundwater level. The relative pavement-life ratios (N_{fn}/N_{fe}) calculated using the lowest N_f (worst case) of the two interface bond conditions (full bond and full slip) and the two distresses (fatigue cracking and rutting) are presented.

Adaptation Cost Analysis

Adaptation costs were calculated for each of the adaptation options. Costs included the NHDOT material, excavation, and construction costs required to enhance the existing structure to achieve the adaptation option. Material costs were estimated by converting the unit costs to US$/sq yd-in. and multiplying this by the thickness of additional or replacement material. Excavation costs were estimated for structures requiring AC removal and base-material replacement. Construction unit costs were estimated using machine operation and productivity values from Valle et al. (*24*). The unit costs (*25*) and the resilient modulus (*17–19*) for the materials used in this analysis are presented in Table 3.

The adaptation strategies meeting the performance threshold for both 1.0 ft (0.3 m) and 2.7 ft (0.8 m) SLR scenarios are presented with the costs to illustrate the most cost-effective adaptation options. The road-surface elevations corresponding with various pavement structures are also discussed with respect to future road-surface inundation.

Discussion of Results

Pavement Service-Life Reduction: Existing Structures

The existing pavement structures and the simulated groundwater levels for the current condition and SLR scenarios are presented in Figure 2. The corresponding pavement service-life reductions caused by SLR-induced groundwater rise are also shown.

Groundwater rises into the pavement base layers at both case-study sites with 1.0 ft (0.3 m) and 2.7 ft (0.8 m) SLR and the road surfaces are projected to flood with

Table 2. Adaptation Options Analyzed for Route 286 (Regional Corridor) and Gosling Road (Local Road) in the Order of Increasing Construction Complexity

	AC	Base 1		Base 2		Subbase	
Description	Thickness (in.)	Thickness (in.)	Material	Thickness (in.)	Material	Thickness (in.)	Material
Route 286							
Existing structure	5.5	16	G	na	na	na	na
Increase AC thickness	6.0	16	G	na	na	na	na
	6.5	16	G	na	na	na	na
	8.0	16	G	na	na	na	na
	8.5	16	G	na	na	na	na
	9.0	16	G	na	na	na	na
Increase G thickness	5.5	20	G	na	na	na	na
	6.0	20	G	na	na	na	na
	5.5	24	G	na	na	na	na
	6.0	24	G	na	na	na	na
	5.5	28	G	na	na	na	na
Add RSB on top of G base	5.5	12	RSB	16	G	na	na
Add CS on top of G base	5.5	8	CS	16	G	na	na
	6.0	8	CS	16	G	na	na
	5.5	12	CS	16	G	na	na
	5.5	16	CS	16	G	na	na
Replace G base CG	5.5	20	CG	na	na	na	na
	5.5	24	CG	na	na	na	na
	5.5	28	CG	na	na	na	na
Replace G base with CS	5.5	20	CS	na	na	na	na
	5.5	24	CS	na	na	na	na
	5.5	28	CS	na	na	na	na
Replace G base with CG, add CS	5.5	12	CS	12	CG	na	na
	5.5	16	CS	12	CG	na	na
Gosling Road							
Existing structure	4.0	12	CG	12	G	8	S
Increase AC, decrease CG base	5.0	11	CG	12	G	8	S
	6.0	10	CG	12	G	8	S
Increase AC, replace CG base and decrease thickness	5.0	11	CS	12	G	8	S
	5.0	11	RSB	12	G	8	S
Keep AC thickness constant, replace base and subbase layers	4.0	12	CS	12	G	8	S
	4.0	12	CS	12	CS	8	S
	4.0	12	CS	12	CS	8	G
	4.0	12	CS	12	CS	8	CS

Note: AC = asphalt concrete; G = gravel; CS = crushed stone; CG = crushed gravel; RSB = reclaimed stabilized base; S = sand; na = not applicable; 1 in. = 25.4 mm.

groundwater inundation (GWI) between the years 2060 and 2090. Rutting controls pavement failure at the Route 286 evaluation site due to the relatively thin gravel base. In contrast, fatigue cracking controls pavement failure at Gosling Road with a thin AC layer atop a more robust base and subbase structure. The controlling distress for pavement failure is indicated by the lowest N_f between rutting and fatigue.

Route 286 is projected to experience approximately 80–90% service-life reduction from rutting by 2030 with 1.0 ft (0.3 m) SLR as groundwater moves into the pavement base layer. Assuming a standard design life of 20 years with appropriate maintenance, this represents a service-life reduction of 16–18 years if adaptation strategies are not implemented. Gosling Road is projected to experience only slightly more than 10% service-life reduction from fatigue cracking with 1.0 ft (0.3 m) SLR increasing to approximately 45% with 2.7 ft (0.8 m) SLR predicted by 2060. Consequently, rising groundwater is projected to reduce the pavement service-life by only 2 years by 2030 and 9 years by 2060 if adaptation strategies are not implemented. Rutting may become more of a factor at Gosling Road if the AC is weakened by rising temperatures associated with climate change.

Road-surface inundation is projected to occur at Route 286 in the mid-to-late 2070s with standing-water depth slightly less than 7 in. (18 cm) predicted with 5.2 ft (1.6 m) SLR by the year 2090 increasing to 13 in. (33 cm) with 6.6

Table 3. Material and Construction Costs for New Hampshire Pavements (*25*) with Resilient Modulus

Description	Unit costs (US$/yd^2-in.)	Resilient modulus (psi)
Material		
Asphalt concrete	3.64	325,000
Crushed stone	0.60	31,000
Reclaimed stabilized base	0.14	31,000
Stone for reclaimed stabilized base	0.83	31,000
Crushed gravel	0.82	21,000
Gravel	0.66	15000
Sand	0.45	9,000
Construction costs		
Asphalt concrete	11.23	
Base and subbase	1.17	
Excavation	0.20	

Note: US$1.00/yd^2-in. = US$0.47/m^2-cm; 1 psi = 6.89 kPa.

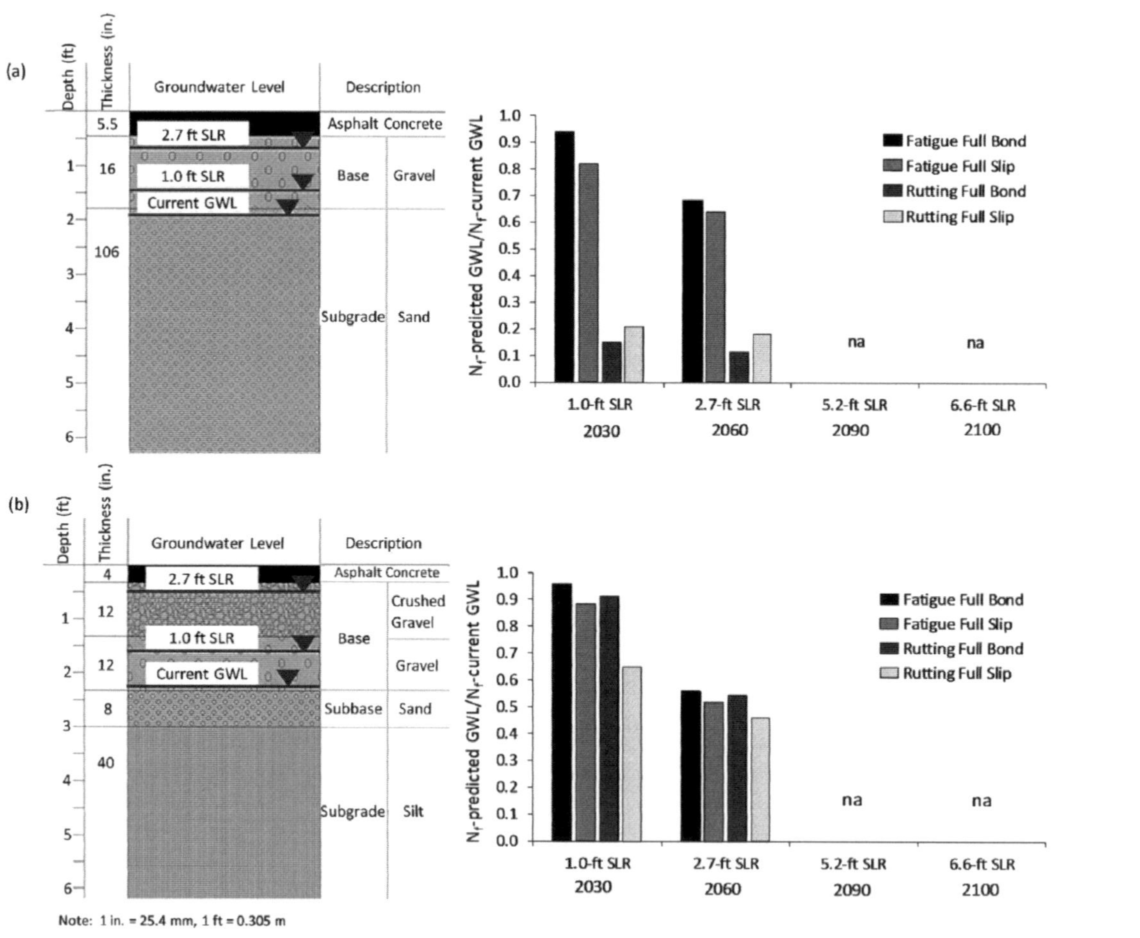

Figure 2. Existing pavement structures to a depth of 6.0 ft (1.8 m) and pavement service-life ratios (N_{fp} / N_{fc}) with SLR-induced groundwater rise: (a) Route 286 (regional corridor) and (b) Gosling Road (local road) (*8*) (SLR = sea-level rise; GWL = groundwater level; na = road surface is flooded).

Table 4. Route 286 Adaptation Options with the Resultant Pavement Service-Life Relative to the Existing Conditions Represented by N_{fn}/N_{fe} for 1.0 ft and 2.7 ft SLR

AC thickness	Base 1: Thickness and material		Base 2: Thickness and material		Road-surface elevation increase	Controlling distress	Resultant pavement life (N_{fn}/N_{fe})		Adaptation costs
(in.)	(in.)	Material	(in.)	Material	(in.)		1.0-ft SLR	2.7-ft SLR	US$/yd^2
Existing structure									
5.5	16	G	0	na	0.0	Rutting	0.15	0.11	0.00
Increase AC layer thickness									
6.0	16	G	0	na	0.5	Rutting	0.20	0.16	7.44
6.5	16	G	0	na	1.0	Rutting	0.28	0.22	14.87
8.0	16	G	0	na	2.5	Rutting	0.69	0.59	37.18
8.5	16	G	0	na	3.0	Rutting	0.92	0.79	44.62
9.0	16	G	0	na	3.5	Rutting	**1.21**	**1.05**	52.06
Increase gravel-base thickness									
5.5	20	G	0	na	4.0	Rutting	0.37	0.28	90.23
6.0	20	G	0	na	4.5	Rutting	0.49	0.37	97.67
5.5	24	G	0	na	8.0	Fatigue	**1.23**	0.83	97.59
6.0	24	G	0	na	8.5	Fatigue	**1.68**	**1.16**	105.02
5.5	28	G	0	na	12.0	Fatigue	**1.40**	**1.00**	104.94
Add reclaimed stabilized base on top of gravel base									
5.5	12	RSB	16	G	12.0	Fatigue	**1.41**	**1.29**	102.93
Add crushed stone on top of gravel base									
5.5	8	CS	16	G	8.0	Fatigue	**1.05**	0.90	97.09
6.0	8	CS	16	G	8.5	Fatigue	**1.42**	**1.24**	104.53
5.5	12	CS	16	G	12.0	Fatigue	**1.41**	**1.29**	104.20
5.5	16	CS	16	G	16.0	Fatigue	**1.86**	**1.77**	111.30
Replace gravel base with crushed gravel									
5.5	20	CG	0	na	4.0	Fatigue	**1.29**	0.81	125.95
5.5	24	CG	0	na	8.0	Fatigue	**1.59**	**1.01**	133.94
5.5	28	CG	0	na	12.0	Fatigue	**1.84**	**1.27**	141.93
Replace gravel base with crushed stone									
5.5	20	CS	0	na	4.0	Fatigue	**1.77**	**1.00**	121.54
5.5	24	CS	0	na	8.0	Fatigue	**2.24**	**1.33**	128.65
5.5	28	CS	0	na	12.0	Fatigue	**2.62**	**1.76**	135.75
Replace gravel base with crushed gravel and add crushed stone									
5.5	12	CS	12	CG	8.0	Fatigue	**2.53**	**1.21**	131.29
5.5	16	CS	12	CG	12.0	Fatigue	**2.84**	**1.67**	138.40

Note: AC = asphalt concrete; G = gravel; CS = crushed stone; CG = gravel; RSB = reclaimed stabilized base; na = not applicable; SLR = sea-level rise; 1 in. = 25.4 mm; 1 ft = 0.305 m. Bold font indicates meets performance threshold.

ft (2.0 m) SLR by end of century. GWI is projected to occur sooner (by the year 2070) at Gosling Road with a standing-water depth on the road surface of approximately 13 in. (33 cm) by 2090 increasing to 21 in. (53 cm) by the year 2100.

In summary, pavement service-life reduction from SLR-induced groundwater rise at the Route 286 site will occur sooner and be more severe than at the Gosling Road site because of the thin base layer in the Route 286 pavement structure. GWI, on the other hand, is projected to occur more than 5 years earlier at the Gosling Road site with a greater standing-water depth than at the Route 286 site.

Adaptation Strategies and Costs

Route 286 (Regional Corridor). Pavement analysis results for the Route 286 adaptation options are presented in Table 4. Only the groundwater levels for 1.0 ft (0.3 m) and 2.7 ft (0.8 m) SLR were considered in this analysis as GWI is projected with 5.2 ft (1.6 m) SLR and beyond. The options are ordered by increasing construction complexity, with simple AC overlays the easiest to implement, and base and subbase material replacement the most complex. Options meeting the performance threshold for both SLR scenarios have the N_f ratios highlighted in bold.

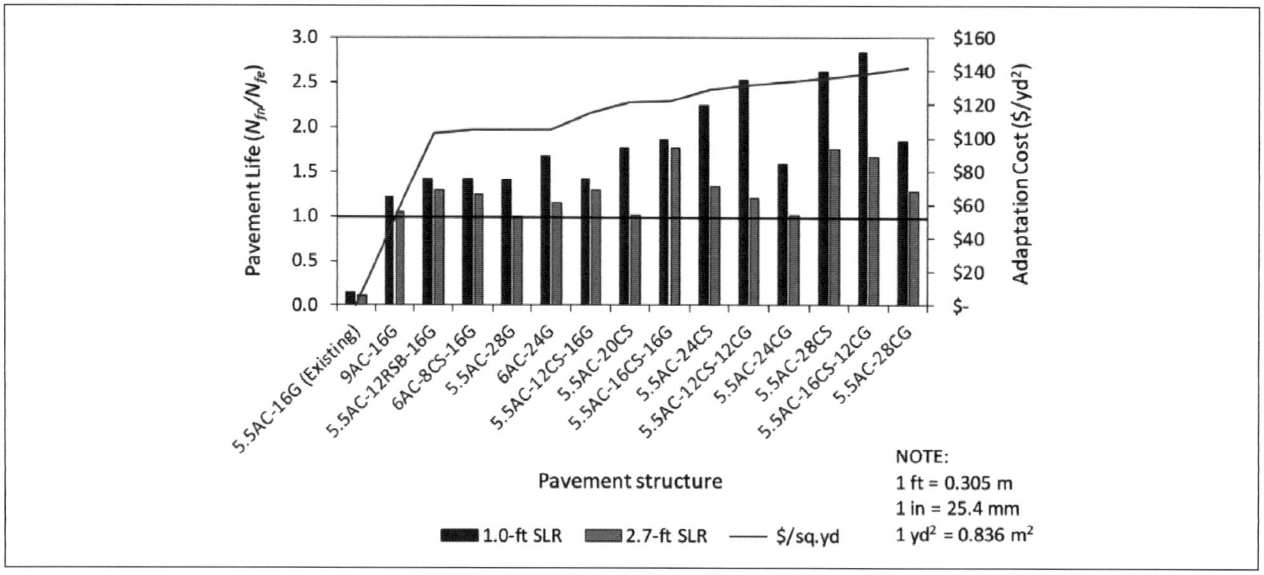

Figure 3. Route 286 (regional corridor) adaptation strategies meeting the performance threshold for 1.0 and 2.7 ft SLR with construction costs (thickness in inches; AC = asphalt concrete; G = gravel; CS = crushed stone; CG = crushed gravel; SLR = sea-level rise; horizontal line = performance threshold).

The distress controlling pavement failure switches from rutting, in the existing structure, to fatigue cracking in all adaptation options with gravel-base thicknesses of more than 20 in. (51 cm). An AC layer thickness of 9.0 in. (23 cm) or more with the existing 16 in. gravel base will provide enough protection for the weakened underlying layers with 1.0 and 2.7 ft (0.3 and 0.8 m) SLR projected to occur by 2030 and 2060, respectively. AC overlays less than 3.5 in. (9 cm) thick (with the existing base structure) will not prevent premature rutting failure with rising groundwater by the year 2030. Pavement structures that meet the performance threshold without increasing the AC thickness include adding 12 in. (30 cm) crushed stone atop the existing 16 in. (41 cm) gravel base or substituting 12 in. (30 cm) RSB supplemented with crushed stone to achieve the same adaptation result. In addition, options with a 28 in. (71 cm) gravel base, 24 in. (61 cm) crushed-gravel base, or 20 in. (51 cm) crushed-stone base will meet the performance threshold for both SLR scenarios.

The road-surface elevation is an important consideration for both case-study sites because the road surfaces are projected to be inundated between 2060 and 2080. Continuous road-surface flooding will render Route 286 useless as an evacuation route unless actions to raise the road are implemented. The increase in road-surface elevation is shown for each option in Table 4. Road-surface elevation increases of 3.5 in. (9 cm) and 12 in. (30 cm) are projected to delay GWI by 8 years and 20 years, respectively. Viable adaptation options with road-surface elevation increases of 12 in. (30 cm) or more are those with granular base layers at least 28 in. (71 cm) thick.

The viable Route 286 adaptation options, that is, those that exceed the performance threshold for both SLR scenarios, are presented with costs in Figure 3. The adaptation costs range from US$52/yd^2 to US$142/yd^2 (US$62/m^2 to US$170/m^2). A large cost increase from US$52/yd^2 to US$103/yd^2 (US$62/m^2 to US$123/m^2) occurs when the underlying base layers require modification and remains relatively constant at US$105/yd^2 (US$126/m^2) for the next four options.

The least expensive option (9AC-16G) adds 3.5 in. (9 cm) of AC overlay to the existing pavement structure. This increases the designed (existing) pavement service-life by 20% with 1.0 ft (0.3 m) SLR and preserves the current service-life until 2060. It adds 3.5 in. (9 cm) additional elevation, delaying road-surface inundation by 8 years if the pavement is maintained and enhanced as groundwater rises. This option will not meet the performance criteria, however, if there is pre-existing rutting damage in the underlying layers. The existing pavement structure at Route 286 is projected to experience large service-life reductions from rutting distress with groundwater rise (Figure 2), so it is essential that adaptation actions be taken well before 2030 to prevent damage to the underlying layers. The current groundwater level at Route 286 is just below the bottom of the base layer suggesting that this adaptation option, if chosen, should be implemented in the next renewal cycle to avoid the increased cost of rehabilitating the base layers.

When considering both the road-surface elevation and pavement life, the most cost-effective adaptation option (5.5AC-12RSB-16G) places 5.5 in. (13 cm) new AC atop 12 in. (30 cm) RSB, all resting on the pre-existing 16 in. (41 cm)

Table 5. Gosling Road Adaptation Options with Resultant Pavement Service-Life Relative to the Existing Condition Represented by N_{fn}/N_{fe} for 1.0 and 2.7 ft SLR

AC	Base 1: Thickness and material		Base 2: Thickness and material		Subbase: thickness and material		Controlling distress	Resultant pavement life (N_{fn}/N_{fe})		Adaptation costs
(in.)	(in.)	Material	(in.)	Material	(in.)	Material		1.0-ft SLR	2.7-ft SLR	US\$/yd^2
Existing structure										
4	12	CG	12	G	8	S	Fatigue	0.88	0.52	0.00
Increase AC layer while decreasing CG base layer										
5	11	CG	12	G	8	S	Fatigue	**1.62**	**1.17**	75.35
6	10	CG	12	G	8	S	Fatigue	**2.98**	**2.49**	90.42
Increase AC layer, replace CG base with stiffer material, and decrease thickness										
5	11	RSB	12	G	8	S	Fatigue	**2.04**	**1.35**	96.95
5	11	CS	12	G	8	S	Fatigue	**2.04**	**1.35**	97.04
Keep AC thickness constant and replace unbound layers with stiffer materials										
4	12	CS	12	G	8	S	Fatigue	**1.23**	0.63	83.94
4	12	CS	12	CS	8	S	Fatigue	**1.33**	0.73	107.61
4	12	CS	12	CS	8	G	Fatigue	**1.38**	0.76	123.88
4	12	CS	12	CS	8	CS	Fatigue	**1.43**	0.79	123.39

Note: AC = asphalt concrete; G = gravel; CG = crushed gravel; RSB = reclaimed stabilized base; CS = crushed stone; S = sand; SLR = sea-level rise; 1 in. = 25.4 mm; 1 ft = 0.305 m. Bold font indicates meets performance threshold.

gravel layer. It increases the designed service-life by 40% and 30% with 1.0 ft and 2.7 ft (0.3 m and 0.8 m) SLR, respectively. In addition, the road-surface elevation is increased 12 in. (30 cm), delaying inundation of the road surface until the year 2100. This option has the best predicted performance of the four least-costly options and costs US\$103/yd^2 (US\$123/m^2), almost doubling the cost of the 9 in (23 cm) AC option. Although expensive, it provides a stronger base, upon which AC overlays can be placed, extending the pavement life even further.

The most robust cost-effective option (5.5AC-16CS-16G) adds 16 in. (41 cm) crushed stone atop the existing 16 in. (41 cm) gravel layer. This option costs US\$122/yd^2 (US\$146/yd^2) and almost doubles the designed service-life with 1.0 ft (0.3 m) SLR and increases it by 80% with 2.7 ft (0.8 m) SLR. It also increases the road-surface elevation by 16 in. (41 cm) making the evacuation route resilient to surface-water flooding and GWI beyond 2100.

Gosling Road (Local Road). The Gosling Road pavement analysis results are presented in Table 5. The pavement surface was constrained to the existing elevation to accommodate adjacent properties. Fatigue-cracking controls the pavement failure for the existing structure and all adaptation options considered at this location. Four options meet the performance threshold for the two SLR scenarios. Increasing the AC layer thickness from 4–5 in. (10–13 cm) while reducing the base-layer thickness to maintain the pavement-surface elevation will provide enough protection for the weakened underlying layers with 1.0 and 2.7 ft (0.3 and 0.8 m) SLR. All pavement structures analyzed with 4 in. (10 cm) AC thickness, despite stiffening the base and subbase layers with crushed stone or crushed gravel, will suffer premature fatigue-cracking failure with rising groundwater between the years 2030 and 2060.

The viable Gosling Road adaptation options and costs are presented in Figure 4. The adaptation costs range from US\$75/yd^2 to US\$97/yd^2 (US\$90/m^2 to US\$116/m^2). The lowest-cost option removes 5 in. (13 cm) of surface material (AC and crushed-gravel base) and repaves with 5 in. (13 cm) of new AC. The underlying base and subbase materials remain untouched. It increases the designed pavement service-life by 60% with 1.0 ft (0.3 m) SLR and by 17% with 2.7 ft (0.8 m) SLR. Replacing the base layers with various thicknesses of stiffer materials while keeping the AC layer thickness constant is more expensive and has no beneficial effect. The most robust adaptation option (6AC-10CG-12G-8S) has a projected pavement service-life almost 3 and 2.5 times the designed service-life at 1.0 ft (0.3 m) SLR and 2.7 ft (0.8 m) SLR, respectively. This option thickens the AC layer from 4 in. to 6 in. (10 cm to 15 cm) while reducing the crushed-gravel layer by 2 in. (5 cm). At US\$90/yd^2 (US\$108/m^2), it costs US\$15/yd^2 (US\$18/m^2) more than the low-cost alternative. If this adaptation option was implemented by 2020, with appropriate maintenance it should not fail from SLR-groundwater rise until the early 2070s when the road surface is projected to be flooded with GWI. The 1.0 ft (0.3 m) SLR annualized adaptation cost for the most robust option (US\$1.52/yd^2-year, US\$1.82/m2-year) is less than the annualized adaptation costs for the lowest-cost option (US\$2.33/yd^2-year, US\$2.79/m2-year) due the projected increase in pavement life.

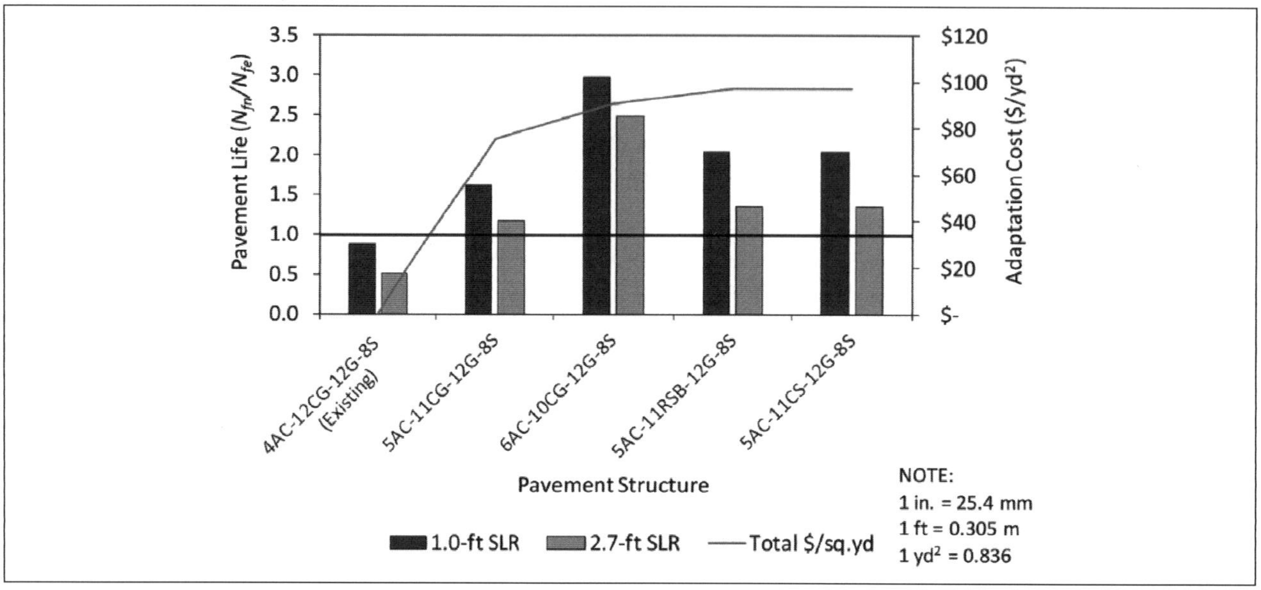

Figure 4. Gosling Road (local road) adaptation strategies meeting the performance threshold for 1.0 ft and 2.7 ft SLR with construction costs (thickness (in.); AC = asphalt concrete; G = gravel; CS = crushed stone; CG = crushed gravel; SLR = sea-level rise; horizontal line = performance threshold).

The analysis of adaptation options for Gosling Road assumes that it is not feasible to increase the road-surface elevation. If this constraint is relaxed, a 1 in., 4 in., and 12 in. (2 cm, 10 cm, and 30 cm) surface-elevation increase would delay GWI of the road surface by 2 years, 6 years, and 18 years, respectively, beyond 2070. The cost of simply adding new asphalt overlays would be lower than the options requiring modification of the base layers, but adjacent property modifications (including a nearby overpass) would add to the overall cost of the adaptation project. Other adaptation options, beyond the scope of this study, might include active groundwater pumping to lower the groundwater table below the bottom of the pavement structure.

Summary and Conclusions

An interdisciplinary approach, coupling groundwater modeling and multi-layer elastic pavement analysis, was used to assess various pavement structural designs that would be resilient in the face of SLR-induced groundwater rise. The groundwater model was used to define the NH-GWRZ and, coupled with lidar ground-surface elevations, used to identify vulnerable roads for four SLR scenarios projected to the year 2100. Approximately 23% (146 miles, 235 km) of roads in NH's Seacoast Region are at risk of premature pavement failure with 6.6 ft (2.0 m) SLR by 2100 if adaptation actions are not taken.

Two case-study sites, predicted to experience pavement service-life reduction, were investigated to determine the effectiveness of different adaptation options. A 2-lane regional connector and evacuation route, located near a coastal wetland, is predicted to experience pavement rutting-life reduction from groundwater rise in the coming decade, reaching 80–90% pavement-life reduction by 2030. Road-surface inundation is projected to occur by the mid-2070s. Adding 12 in. (30 cm) of RSB atop the existing 16 in. (41 cm) gravel base was found to be the most cost-effective adaptation option to mitigate service-life reduction at this location. The 12 in. (30 cm) road-surface elevation increase delays pavement-surface flooding by approximately 20 years to the end of the century. Applying AC overlays less than 3.5 in. (9 cm) thick in response to pavement distress will not protect this pavement structure from premature failure caused by SLR-induced groundwater rise.

An important 4-lane local road, providing access to a shopping mall and international airport is predicted to experience less service-life reduction, 10% by 2030 and 45% by 2060. Fatigue cracking, in contrast to rutting, controls pavement failure at this evaluation site because of a more robust underlying pavement structure; consequently, simply thickening the AC layer on top of the existing base layers is the most cost-effective adaptation option. Adjacent properties may render road-surface elevation increases infeasible at this location, in which case, it would be possible to maintain the designed service-life with rising groundwater by increasing the AC layer 1–2 in. (3–5 cm) with a corresponding reduction in crushed-gravel base thickness. The surface of the pavement is projected to be inundated with rising groundwater by the early 2070s.

Significant reductions in pavement service-life with SLR-induced groundwater rise are predicted if no adaptation strategies are implemented. These reductions will potentially be more severe as the AC weakens with temperature increases caused by climate change. This investigation shows that when base layers are sufficient in material and thickness to protect the subgrade from rutting failure, increasing the AC layer thickness is sufficient to avoid the costs of premature failure from SLR-induced groundwater rise. Coastal roads with inadequate base-layer thicknesses may experience premature rutting failure. It is recommended that the rehabilitation of these roads include projected SLR-induced groundwater rise and pavement temperature changes in design calculations to avoid the high costs of more frequent pavement failures. The adaptation options analyzed here were based on input from NHDOT and local planners, but it is unclear whether the recommendations will be implemented. Considerations in addition to cost and performance, such as base-material availability and design practicality may also factor into implementation decisions.

Future work includes investigating the combined effect of climate-change-induced groundwater rise and increasing pavement temperatures on pavement service-life over a broader range of climate scenarios. Future adaptation work will examine a staged response to groundwater and temperature rise in which a series of shorter-term actions, such as rebuilding a more robust base, are implemented during major road maintenance projects, and thicker AC overlays are applied in future steps.

Acknowledgments

The authors would like to acknowledge NH Sea Grant for funding the work; NH Seacoast Transportation Climate Working Group; NH Department of Transportation (NHDOT); NH Department of Environmental Services (NHDES); the Infrastructure and Climate Network (ICNET), and University of New Hampshire Center for Infrastructure Resilience to Climate (UCIRC). Support for the Infrastructure and Climate Network (ICNet) project was provided by the U.S. National Science Foundation.

References

1. Melillo, J., T. Richmond, and G. Yohe. *Climate Change Impacts in the United States: The Third National Climate Assessment. Chapter 1: Overview and Report Findings*. US Global Change Research Program, 2014.
2. Wu, S., R. Najjar, and J. Siewert. Potential Impacts of Sea-Level Rise on the Mid- and Upper-Atlantic Region of the United States. *Climatic Change*, Vol. 95, Nos. 1-2, 2009, pp. 121–138. https://dx.doi.org/10.1007/s10584-008-9522-x.
3. Johnston, A., P. Slovinsky, and K. L. Yates. Assessing the Vulnerability of Coastal Infrastructure to Sea Level Rise using Multi-Criteria Analysis in Scarborough, Maine (USA). *Ocean & Coastal Management*, Vol. 95, 2014, pp. 176–188. https://dx.doi.org/10.1016/j.ocecoaman.2014.04.016.
4. Masterson, J. P., M. N. Fienen, E. R. Thieler, D. B. Gesch, B. T. Gutierrez, and N. G. Plant. Effects of Sea-Level Rise on Barrier Island Groundwater System Dynamics – Ecohydrological Implications. *Ecohydrology*, Vol. 7, No. 3, 2014, pp. 1064–1071. https://dx.doi.org/10.1002/eco.1442.
5. Walter, D. A., T. D. McCobb, J. P. Masterson, and M. N. Fienen. Potential Effects of Sea-Level Rise on the Depth to Saturated Sediments of the Sagamore and Monomoy Flow Lenses on Cape Cod, Massachusetts. *U.S. Geological Survey Scientific Investigations Report 2016–5058*. U.S. Geological Survey, Reston, VA, 2016.
6. Ping, W., B. Sheng, C. Ling, B. Dietrich, and D. Horhota. Influence of Base Clearance on Subgrade Resilient Modulus of Florida Roadway Pavements. *Transportation Research Record: Journal of the Transportation Research Board*, 2010. 2154: 176–186.
7. Roshani, A. Road Infrastructure Vulnerability to Groundwater Table Variation due to Sea Level Rise. MS thesis. Queensland University of Technology, Australia, 2014.
8. Knott, J. F., J. S. Daniel, J. M. Jacobs, P. Kirshen, and M. Elshaer. Assessing the Effects of Rising Groundwater from Sea-Level Rise on the Service Life of Pavements in Coastal Road Infrastructure. *Transportation Research Record: Journal of Transportation Research Board*, 2017. 2639: 1–10.
9. Kirshen, P., L. Capito, R. M. Vogel, P. Mathisen, A. Rosner, and T. Renaud. Adapting Urban Infrastructure to Climate Change: A Drainage Case Study. *Journal of Water Resources Planning Management*, Vol. 141, No. 4, 2014. http://dx.doi.org/10.1061/(ASCE)WR.1943-5452.0000443.
10. Salour, F., S. Erlingsson, and C. Zapada. Model for Seasonal Variation of Resilient Modulus in Silty Sand Subgrade Soil, Evaluation with Falling Weight Deflectometer. *Transportation Research Record: Journal of the Transportation Research Board*, 2015. 2510: 65–73.
11. Elshaer, M., M. Ghayoomi, and J. Daniel. Methodology to Evaluate Performance of Pavement Structure using Soil Moisture Profile. *Journal of Road Materials and Pavement Design*, 2016, pp.1–20. https://dx.doi.org/10.1080/14680629.2017.1283356.
12. Mallick, R., and T. El-Korchi. *Pavement Engineering Principals and Practice*, 2nd ed. CRC Press, Taylor and Francis Group, LLC, Boca Raton, FL, 2013.
13. Mack, T. J. Assessment of Ground-Water Resources in the Seacoast Region of New Hampshire. *U.S. Geological Survey Scientific Investigations Report*, Vol. 2008–5222, 2009.
14. Harbaugh, A. MODFLOW-2005, the U.S. Geological Survey Modular Ground-Water Model—the Ground-Water Flow Process. *U.S. Geological Survey Techniques and Methods*, Vol. 6-A16, 2005.
15. Parris, A., P. Bromirski, V. Burkett, D. Cayan, M. Culver, J. Hall, R. Horton, K. Knutti, R. Moss, J. Obeysekera, A. Sallenger, and J. Weiss. *Global Sea Level Rise Scenarios for the US National Climate Assesment*. Technical Memo OAR CPO-1, National Oceanic and Atmospheric Administration, Silver Spring, MD, 2012.
16. New Hampshire Department of Transportation (NHDOT). Project Viewer. http://gis.dot.nh.gov/projectviewer/. Accessed May 19, 2016.
17. Christopher, B. R., C. Schwartz, and R. Boudreau. *Geotechnical Aspects of Pavements*. FHWA NHI-05-037, National Highway Institute, Federal Highway Administration, 2006.

18. Janoo, V. *Layer Coefficients for NHDOT Pavement Materials*. Special Report 94-30, U.S. Army Corps of Engineers, Cold Regions Research and Engineering Laboratory (CRREL), Hanover, NH, 1994.
19. New Hampshire Department of Transportation (NHDOT). *Highway Design Manual*. Appendix 7-1 Flexible Pavement Analysis. New Hampshire Department of Transportation, Concord, NH, 2014.
20. Janoo, V., J. J. Bayer, G. D. Durell, and C. E. Smith. *Resilient Modulus for New Hampshire Subgrade Soils for use in Mechanistic AASHTO Design*. Special Report 99-14, U.S. Army Corps of Engineers, Cold Regions Research and Engineering Laboratory (CRREL), Hanover, NH, 1999.
21. AASHTO. *Mechanistic-Empirical Pavement Design Guide (MEPDG). A Manual of Practice, Interim Edition*. Washington, D.C., 2008.
22. Huang, Y. *KENPAVE – A Computer Package for Pavement Analysis and Design*. University of Kentucky, Lexington, KY, 2003.
23. Timm, D., D. Newcomb, and B. Birgisson. *Mechanistic-Empirical Flexible Pavement Thickness Design: The Minnesota Method*. MN/RC-P99-10, MNDOT, St. Paul, MN, 1999.
24. Valle, O., Y. Qiao, E. Dave, and W. Mo. Life Cycle Assessment of Pavements Under a Changing Climate. *Proc., Symposium on Life-Cycle Assessment of Pavements (Pavement LCA 2017)*, Champaign, Illinois, USA, 2017, pp. 241-250.
25. New Hampshire Department of Transportation (NHDOT). *Weighted Average Unit Prices for Projects between 4/1/2016 and 3/31/20*. New Hampshire Department of Transportation, Concord, NH, 2017.

The Special Task Force on Climate Change and Energy (A0020T) peer-reviewed this paper (18-03258).

Article

TRR
JOURNAL OF THE TRANSPORTATION RESEARCH BOARD

Transportation Research Record
2018, Vol. 2672(2) 23–32

DOI: 10.1177/0361198118756894
journals.sagepub.com/home/trr
SAGE

Dealing with the Gap between Type-Approval and In-Use Light Duty Vehicles Fuel Consumption and CO_2 Emissions: Present Situation and Future Perspective

Jelica Pavlovic[1], Konstantinos Anagnostopoulos[1], Michael Clairotte[1], Vincenzo Arcidiacono[1], Georgios Fontaras[1], Iker Prado Rujas[1], Victor Valverde Morales[1], and Biagio Ciuffo[1]

Abstract

There is increasing evidence suggesting that real-world fuel consumption and CO_2 improvements in the last decade have been much less than those measured during type-approval tests. Scientific studies have found that the offset between officially reported values and real-world vehicle CO_2 emissions in Europe has constantly increased over the last years. The difference between officially reported and actual CO_2 emissions of vehicles has three main implications: (i) it undermines the effectiveness of CO_2 regulations in reducing greenhouse gas emissions in Europe; (ii) it distorts competition between vehicle manufacturers; (iii) it undermines innovation. As a fundamental step to deal with this issue, the European Commission has already replaced the old and outdated test procedure used so far in the emission type-approval of vehicles with the worldwide harmonized light vehicles test procedure (WLTP). Being a lab-based test procedure, the WLTP, by its nature, can only cover part of the CO_2 gap. There is therefore increasing pressure to integrate the current type-approval system with additional measures based on real-world measurements. One of the options under discussion is to use the CO_2 emissions measured during the real driving emission test. The objective of the present paper is to assess the validity of this proposal and to propose other possible ways to deal with the CO_2/fuel consumption gap. In particular, the paper presents experimental evidence on the variability of the CO_2/fuel consumption of a vehicle, questioning the idea that a single central estimate of these quantities may be sufficient.

In 2009, the European Union (EU) set a target for the average CO_2 emissions of new passenger cars at 95 g CO_2/km, which applies from the year 2021 on. The Commission is now preparing a review of the legislation with a view of setting new CO_2 targets for light duty vehicles for the next decade.

There is, however, increasing evidence (Table 1 [*1–11*]) that suggests that real-world fuel consumption and CO_2 improvements in the last decade have been much less than those measured during type-approval. Scientific studies have found that the offset between officially reported values and real-world vehicle CO_2 emissions increases year by year, and was estimated at about 36% for 2015. Results from real driving emissions (RDE) tests for CO_2 emissions show differences of the same order (30±12%). Furthermore, several investigations have shown that the reported CO_2 values could not be reproduced under the official laboratory test conditions and that offsets in CO_2 emissions of about 10–15% on average exist (*12*). The situation is not peculiar to the European case. In other research the authors analyzed a database of 75,000 fuel consumption estimates in the U.S. and found that there is a certain gap between test-cycle and in-use fuel economy estimates, and that this gap has been increasing since 2005 (*13*).

However, the gap does not necessarily represents a problem if (a) it remains constant over time, and (b) it remains constant among the fleet (namely it is the same for all vehicles). In fact, the type-approval fuel consumption and CO_2 emission values are usually used for three main purposes: (i) checking vehicle manufacturers' compliance with respect to the CO_2 targets in place in several political contexts; (ii) defining the taxes to be paid for the possession of a specific vehicle; and (iii) allowing drivers to select the right vehicle

[1]Directorate for Energy Transport and Climate, European Commission – Joint Research Centre, Ispra (VA), Italy

Corresponding Author:
Address correspondence to Biagio Ciuffo: biagio.ciuffo@ec.europa.eu

Table 1. Evolution of the Average Gap between Real-World and Type-Approval CO_2 Emissions in Europe Reported from Different Authors in the Period 2011–2016

Authors	Year	Country	Real world – Certification value CO_2 shortfall
Weiss et al. (*1*)	2011	EU	21%
Mellios et al. (2)	2011	EU	25%
Fontaras and Dilara (*3*)	2012	EU	22.5%
Ligterink (*4*)	2013	Netherlands	30%
Mock et al. (*5*)	2014	EU	38%
Ligterink and Eijik (*6*)	2014	Netherlands	44%
Reynaert and Sallee (*7*)	2014	U.S.	41%
Tietge et al. (*8*)	2015	EU	40%
Zacharof et al. (*9*)	2015	EU	30%
Fontaras et al. (*10*)	2016	EU	40%
Duarte et al. (*11*)	2016	EU	24%

to buy (in case fuel consumption is one of the parameters influencing choice). If the gap remains constant over time and it does not depend on the specific vehicle (or the specific technology configuration), the type-approval value of CO_2 and fuel consumption can still be used for these three purposes. The main problem today, especially in light of the increasing number of technology configurations of vehicles entering the market, is that the gap is not only significant, but it is increasing over time and is very different among the different types of vehicles.

There are three main reasons why a widening and heterogeneous gap between officially reported and actual CO_2 emissions of vehicles constitutes a problem: (i) it undermines the effectiveness of the CO_2 regulations for cars and vans in reducing greenhouse gas emissions in Europe; (ii) it distorts competition between vehicle manufacturers; and (iii) it undermines innovation by hampering the development of new vehicle efficiency technology.

Certain initiatives have been already taken to limit the increasing divergence between the officially declared and actual CO_2 emissions of light duty vehicles. The introduction of the new Worldwide harmonized Light vehicles Test Procedure (WLTP) in 2017 will bring type-approval results closer to reality. Through the correlation and target translation legislation (*14,15*) the CO_2 target value will already be linked to the WLTP by 2017, which is an important step toward the definition of new, WLTP-based CO_2 targets for the post-2021 period (*16*). Although RDE testing cannot replace the certification test because of the variability of the test results, it can potentially serve as an additional valuable source of data on the performance of vehicles in terms of CO_2 emissions and fuel consumption under more realistic operating conditions. The usefulness of such data, however, depends on how well the "normal-operating" conditions defined by the RDE test procedure are representative of the variable real-life vehicle operations.

Many studies have tried to provide a thorough explanation of the elements affecting the fuel consumption variability of vehicles in real driving conditions [a recent comprehensive overview has been provided in Fontaras et al. (*10*)]. However, only a few attempts to experimentally determine the possible fuel consumption variability caused by a combination of, for example, temperature/weather, driver, traffic, landscape, road conditions, etc., exist in the literature, and usually only focus on certain aspects of the problem.

As an example, Pathak and co-authors (*17*) investigated the effect of two different roads on the fuel consumption of the same vehicle, concluding that better road management can have a significant influence on fuel consumption (although the scope of the analysis was very limited). A similar approach, although using a significantly more robust experimental set-up and statistical treatment of the results, is proposed by Frey and co-authors (*18*) and by Li and co-authors (*19*). In the first paper three vehicles were used to collect emissions and fuel consumption data on two different origin/destination pairs, each of which was connected by three alternative routes. In the second paper, fuel consumption and pollutant emissions from one vehicle were collected in different road types, traffic, and environmental conditions. In both cases, results showed that the largest effect was due traffic conditions, followed by the road characteristics.

It is worth noting that all the aforementioned studies focused on elements affecting the variability in fuel consumption and emissions, but none tried to quantify the overall variability. This can be considered plausible as all the possible combinations of the relevant factors may cause a very high variability in the fuel consumption of a vehicle. This consideration should already question the possibility that a single test, although carried out in real-life conditions, may be considered representative of the fuel consumption of a vehicle. However, an empirical analysis confirming this consideration is still missing.

The present study therefore aims to fill this gap. In particular, it will present the results of the empirical evidence collected on the subject by the Join Research Centre (JRC), Ispra, Italy. As mentioned, the subject of the study is the characterization of the uncertainty (variability) in vehicle fuel consumption to understand whether an approach based on the determination of a single figure (determined through an ad-hoc RDE test or through corrections to the lab-based type-approval value) can be considered acceptable from a conceptual point of view or not. The paper is organized as follows. In the next section the methodology used for this analysis is briefly presented, followed by a description of the experiments carried out, and by the presentation of the main results achieved. A discussion section then follows to outline the main implications of the outcomes of the study. A final section then summarizes the main the conclusions.

Methodology

The characterization of the uncertainty/variability in the fuel consumption of light duty vehicles was performed using two different data sets:

- Data set no. 1: Data collected during a period of 6 months from a vehicle driven by different drivers for trips with different origin and destination and with different environmental conditions;
- Data set no. 2: Data collected from the RDE tests carried out at the JRC during recent years using different vehicles with a few drivers and considering a limited number of routes.

The two data sources have different characteristics. In the first experiment, everything about the vehicle usage, the trips, and the drivers is known, and therefore an in-depth characterization of fuel consumption variability through both quantification of the uncertainty and understanding the main factors leading to this is possible. However, the analysis refers to a single vehicle only, and therefore the results are certainly biased by its technology configuration. The second experiment includes different vehicles. Much information about the vehicles and the routes is known. The number of tests per vehicle is limited (around 10 trips per vehicle), and there are a limited number of drivers and routes. Some characterization of the uncertainty is possible, but the reasons leading to it are not necessarily captured.

In spite of their intrinsic limitations, the use of these two sets of data in the same study allows validation of one with the other, and enables the results to be extended to achieve dimensions that would not be possible if the two studies were taken in isolation.

The two data sets required different approaches to quantify the fuel consumption variability. In the first case, as only one vehicle is considered, the specific fuel consumption (in l/100 km) can be directly used in the analysis. In the second case, as different vehicles are involved, a direct comparison of the fuel consumption is not possible. However, as the focus of the study is not the absolute fuel consumption but rather its variability, all the fuel consumption values have been normalized with respect to the average fuel consumption measured for that vehicle throughout the different RDE tests. In addition, as each RDE test requires driving the vehicle in three different conditions (urban, rural, and motorway), all three fuel consumption values plus their combination were considered to enrich the information content of the entire sample. In this picture, two normalizations were also applied: a first normalization based on the average CO_2 emission over the corresponding phase (Urban, Rural, Motorway, and Combined) for each vehicle (Norm_1), and a second normalization based on the average CO_2 emission of the vehicle, regardless of the phase (Norm_2).

Table 2. Specification of the Vehicle and Details Related to Tests in the Laboratory and on the Road

Vehicle Class	M1
Model Year	2016
Engine capacity/power	1956 cm^3/103 kW
Transmission	Automatic (9-speed)
Emission standard	Euro6b
After-treatment system	DOC, EGR (low and high pressure), LNT, DPF
RL coefficients NEDC (JRC estimations)	Inertia: 1590 kg; F0: 101; F1: 0.351; F2: 0.034
Number of drivers	8
Number of trips	197
Number of segments	217
Total mileage driven (km)	6716

It is worth mentioning here that for the RDE tests, the analysis is based on CO_2 and not on fuel consumption because fuel consumption is not measured during these tests. However, given the approximately constant relationship between CO_2 and fuel consumption, using one or the other has no effect on the analysis of their variability.

The analysis was carried out by plotting the histograms of the fuel consumption distributions derived from the two samples and by computing the main statistics from the two distributions.

Case Study

Data Set No. 1

Data set no. 1 was obtained by collecting driving data from different drivers using the same vehicle for a period of 6 months to satisfy their driving needs. For this, a vehicle was hired for a period of 12 months. The vehicle was equipped with an On-Board Diagnostics (OBD) II logger ad-hoc developed to access all the information available at the electronic control unit of the specific vehicle. In addition, the device was complemented by a GPS system (to monitor the position of the vehicle over time), and by two current-measurement systems to monitor the operation of the battery and of the alternator. In this way, it was possible to achieve a complete characterization of the vehicle operations.

The vehicle chosen for the analysis was a 2.0 l C segment diesel vehicle equipped with automatic transmission (see Table 2 for details). The vehicle is compliant with Euro 6 regulation for pollutant emissions. Automatic transmission reduces the potential variability in fuel consumption, as the gear-shift behavior of the users can also potentially affect fuel consumption and emissions. The vehicle has three fuel consumption-relevant driving modes (normal, sport, and 4 by 4 mode), and the selection of modes was left to the drivers. The vehicle also has engine start/stop function that was used for most trips.

Before running on-road tests, the vehicle was tested in the Vehicle Emissions LAboratories (VELA) of the European Commission JRC. The VELA test cell used is equipped with a four-wheel drive (4WD) chassis dynamometer with two roller benches of diameter 48 in, maximum traction force: 3300 Nm, inertia range: [454, 2720] kg and allowing vehicle soaking within a temperature range [–30, +50]°C. The test was carried out following the New European Driving Cycle (NEDC) test procedures. Inertia and road load coefficients applied for NEDC testing have been estimated by the authors, as the official values are not available, and can be found in Table 1. In particular, an algorithm developed at JRC able to correlate road loads with the main physical characteristics of the vehicles was used. The NEDC tests were done in three repetitions.

After the tests carried out in the VELA, the vehicle underwent RDE tests following the procedure defined in the legislation recently adopted by the European Commission (*20*). The objective of the RDE tests is the measurement of pollutant emissions (NOx, CO, PM/PN) under realistic in-use conditions. The measurement is carried out using a portable emissions measurement system (PEMS). For the present study, the vehicle was also equipped with the ad-hoc developed OBD logger during these tests.

After the two test operations were completed, the vehicle was provided free of charge to eight different drivers, each of them using the vehicle for a period of 2 weeks. The drivers were selected among JRC colleagues who expressed an interest in taking part in the testing. On accepting the vehicle, the drivers signed a declaration of intent in which they committed to use the vehicle as they would do their own.

Fuel was not provided, so as not to influence the driving style. Driving activities for the first couple of trips of data acquisition were discarded and not taken into account in the analysis, to allow at least a basic familiarization with the vehicle. Drivers were requested to take note of the following elements at the end of each driving activity: weather conditions, traffic conditions, number of passengers, specific additional loads added to the vehicle (luggage, etc.), and general functioning of the vehicle, as well as any unexpected situation faced during the trip. All drivers were employees of the European Commission JRC for insurance-related reasons. In travel-related surveys/experiments, income and instruction levels are usually taken into consideration as sample stratification dimensions; thus the sample of people driving the vehicle might show a polarization with respect to these two dimensions. This is not considered as a strong limitation as this induces a slightly reduced uncertainty in the measured fuel consumption that, given the objective of the paper, can be considered as a safety margin. Attention was paid to have a wide range of trip characteristics, trying to cover urban, rural, motorway, and mixed driving conditions. However, because of the configuration of the local area, there is a slightly higher prevalence of rural driving conditions. At the same time, because of the morphology of the area, the influence of slope and road geometry is expected to contribute to an increase in uncertainty in the fuel consumption.

Tests evaluated in the present paper were recorded during the period between December 2016 and April 2017. The number of trips recorded during this period is 197, and in total 6716 km were driven (please refer to Table 1). All trips containing pauses longer than 5 minutes were split into different trips. The minimum duration of a trip was set to 5 minutes. In total, 217 trips were derived and used to draw the distribution of the vehicle fuel consumption. Several OBD signals (between 80 and 90) were registered during the trips. However, for this report the relevant signals are vehicle speed (km/h) and fuel consumption (l/h), from which the average fuel consumption (l/100km) of each segment is calculated.

Data Set No. 2

The second data set comprises the fuel consumption experienced by 37 vehicles undergoing a total of 274 RDE tests in the period 2014–2016. On the basis of the characteristics of the trip, each is divided, according to the RDE test procedure, into urban, rural, and motorway conditions. The three conditions are defined as follows: (i) urban operations are characterized by vehicle speeds up to 60 km/h; (ii) rural operations by vehicle speed between 60 and 90 km/h; and (iii) motorway operation by vehicle speeds higher than 90 km/h. The CO_2 measured during the trip divided by the total distance traveled represents the combined CO_2. Performing this differentiation, from the 274 RDE tests, 547 different values of fuel consumption/CO_2 emissions were derived to assess the uncertainty in fuel consumption values. As the 547 measures of consumption refer to 37 different vehicles (differing in technology configuration, size, emission legislation, etc.), the statistics could not be directly drawn on the absolute CO_2 emissions and a normalization of the CO_2 was necessary.

To clarify how this was carried out, Table 3 reports the elaborations carried out on the data from the first vehicle tested. From the table it can be seen that the vehicle underwent five RDE tests. In two cases the tests did not include motorway driving. For this reason, 18 CO_2 observations were derived. Two types of normalization were applied. In the first normalization (Norm_1) each CO_2 value is normalized by the average CO_2 emissions of the vehicle under the same driving conditions. As an example, values reported in the fifth row of the table (Norm_1 CO2 Urban) have been obtained by dividing the CO_2 emissions reported in the first row (CO2 Urban) by the average of the CO_2 emissions in the same row. The second normalization (Norm_2) is carried out with respect to the average of all the CO_2 values from the same vehicle. Similar to the previous example, values reported in the ninth row of the table (Norm_2 CO2 Urban) were obtained by dividing the CO_2 emissions reported in the

Table 3. Elaboration of the Results of the RDE Tests of Vehicle No. 1

		RDE trip				
Vehicle no. 1	Unit	1	2	3	4	5
CO_2 Urban	g/km	217.32	199.35	176.84	169.18	195.60
CO_2 Rural	g/km	126.54	143.83	137.96	179.48	87.27
CO_2 Motorway	g/km	165.22	176.21	168.83	–	–
CO_2 Combined	g/km	173.49	175.49	164.27	169.49	190.28
Norm_1 CO_2 Urban	–	1.13	1.04	0.92	0.88	1.02
Norm_1 CO_2 Rural	–	0.94	1.07	1.02	1.33	0.65
Norm_1 CO_2 Motorway	–	0.97	1.04	0.99	–	–
Norm_1 CO_2 Combined	–	0.99	1.01	0.94	0.97	1.09
Norm_2 CO_2 Urban	–	1.30	1.19	1.06	1.01	1.17
Norm_2 CO_2 Rural	–	0.76	0.86	0.82	1.07	0.52
Norm_2 CO_2 Motorway	–	0.99	1.05	1.01	–	–
Norm_2 CO_2 Combined	–	1.04	1.05	0.98	1.01	1.14

Table 4. Comparison of Phase and Combined Fuel Consumption (l/100 km) NEDC Type-Approval and Test Result Values

Test	Phase 1: UDC	Phase 2: EUDC	Combined
Type-Approval NEDC	6.5	4.9	5.5
Measured NEDC	7.7	5.4	6.2

first row of the table (CO2 Urban) by the average of the 18 different values of CO_2 emissions.

Results

The present study tries to draw inference about the uncertainty in the vehicles' fuel consumption from two different sets of experimental data. The results are reported in the following sections.

Data Set No. 1

Table 4 shows average fuel consumption per phase and combined (the NEDC test procedure uses a cycle which is divided into two phases, representing urban and extra-urban driving conditions) together with the official value reported by the vehicle (type-approval value). As can be seen from Table 4, the combined test results were higher (6.2 l/100 km) than the declared value (5.5 l/100km). Accordingly, the phase (Urban Driving Cycle: UDC and Extra-Urban Driving Cycle: EUDC) test results were also higher than the corresponding type-approval values. The difference of approximately 13% (declared vs. measured) may be a series of procedural elements intrinsic to the NEDC procedure [for a complete overview, please refer to Pavlovic et al (*21*)]. Such a deviation is in line with previous experience reported by the authors (*12*).

Table 5 reports the fuel consumption measured during the RDE tests. It is also reported in the table whether the RDE test was a valid test or not. A valid test fulfills the criteria contained in the European RDE legislation (*20*) concerning urban, rural, motorway distance shares, driving dynamics, etc. In particular, Table 5 reports the specific fuel consumption (l/100km) for each RDE test and the statistical mean value on all trips, on all the compliant trips and on all the non-compliant trips. The statistical coefficient of variation for each of the three sets of data is also reported next to the average value (the coefficient of variation is the ratio between the mean and the standard deviation of a distribution).

From the results of the laboratory and RDE tests it is possible to draw the following considerations:

(1) As expected, the fuel consumption from all RDE trips is always higher than the corresponding NEDC values. However,

- considering the average of *all RDE trips* and the type-approval value for the vehicle, the gap is approximately 33%;
- considering the average of *all RDE trips* and the measured NEDC value, the gap is approximately 18%;
- considering the average of *all RDE valid trips* and the type-approval value, the gap is approximately 29%; and
- considering the average of *all RDE valid trips* and the measured NEDC value, the gap is approximately 15%.

Even though the results refer to a single vehicle, considering the figures reported in Table 1 for the average gap derived from different studies, it can be argued that CO_2 emissions and fuel consumption measured during valid RDE trips would be able to cover only part of the existing gap. As,

Table 5. Fuel Consumption (l/100 km) of PEMS Trips (Complete Trip and Phase Results)

PEMS Test ID	Valid	Complete	Urban	Rural	Motorway
1	No	7.0	8.1	5.6	6.8
2	No	7.4	8.2	5.3	8.4
3	Yes	7.1	7.9	5.4	7.8
4	Yes	7.3	8.7	5.7	7.2
5	No	7.3	8.1	5.3	8.2
6	No	7.9	9.2	6.0	8.1
7	No	7.9	9.0	5.9	8.3
8	Yes	7.1	8.3	5.4	7.2
9	Yes	7.0	8.2	5.7	7.0
10	Yes	7.1	8.0	5.3	7.6
Average of compliant tests		7.1±1.5%	8.2±3.8%	5.5±3.4%	7.4±4.5%
Average of non-compliant tests		7.5±5.2%	8.5±6.3%	5.6±5.8%	8.0±8.3%
Average of all PEMS tests		7.3±4.6%	8.4±5.3%	5.6±4.7%	7.7±7.6%

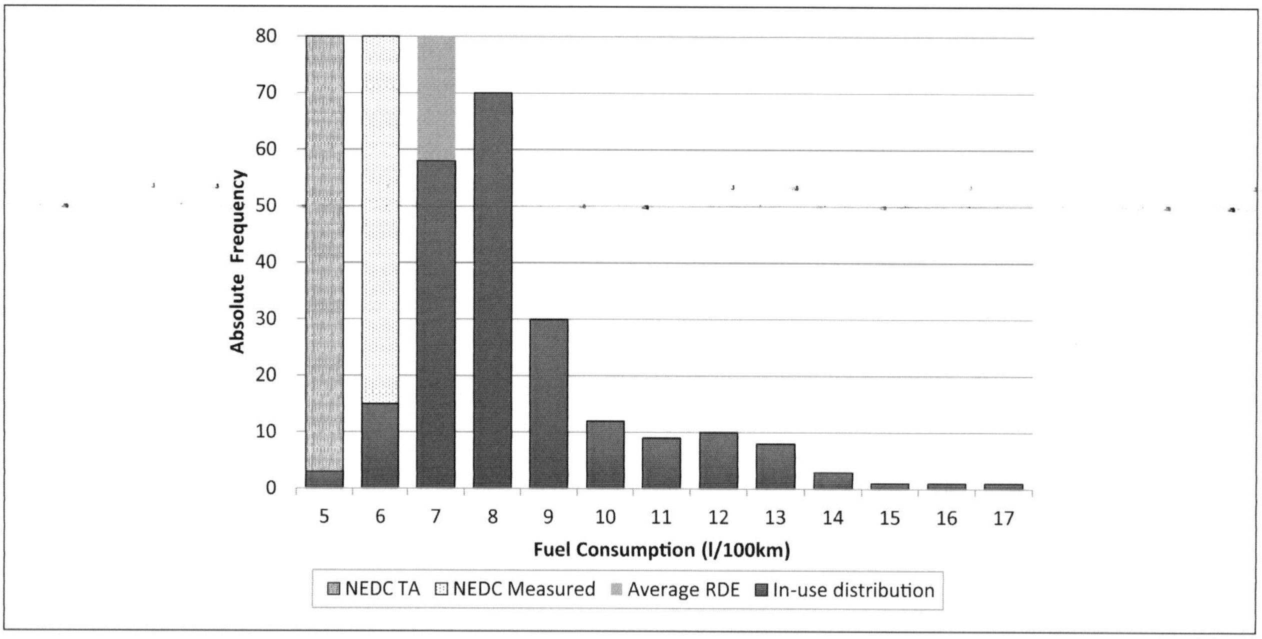

Figure 1. Fuel consumption distribution from all trips collected.

according to some studies [see, for example, Tsiakmakis et al. (*22*)] the introduction of the WLTP (*23*) will already be able to increase the average type-approval fuel consumption of approximately 20%, it does not seem necessary to introduce at this stage a new value which in any case is not able to cover the entire gap.

(2) The fuel consumption derived from valid RDE trips is systematically lower and less variable than the fuel consumption derived from non-valid trips. This tendency is interesting for two reasons: the constraints introduced by the legislation limit the variability of the results, and this is likely to increase the repeatability of such a procedure (although the tests carried out in the present study only considered a very limited number of routes and conditions). At the same time, however, it shows that the fuel consumption from an RDE trip it is likely to represent a rather conservative estimation of the vehicle fuel consumption. This is confirmed by the results reported in Figure 1 on the distribution of the fuel consumption from different drivers.

Figure 1 shows the distribution of fuel consumption from all the trips collected by the eight volunteer drivers. For comparison purposes, the figure also reports the type-approval and measured NEDC fuel consumption, as well as the average fuel consumption from the different RDE tests.

Descriptive statistics related to Figure 1 can be found in Table 6, also for each single driver. The average fuel consumption for each driver was in the range from 6.9 l/100 km (Driver 1) to 9.8 l/100 km (Driver 7); this is significantly

Table 6. Descriptive Statistics for Fuel Consumption of Segments for Each Driver and All Segments Together

Statistical test	All trips	Driver 1	Driver 2	Driver 3	Driver 4	Driver 5	Driver 6	Driver 7	Driver 8
Mean[a]	8.0	6.9	9.5	7.7	7.5	7.7	7.2	9.8	8.9
Median[a]	7.4	6.7	9.0	7.3	7.0	7.5	7.1	8.7	7.9
St. Dev.[a]	2.1	1.4	2.5	1.7	1.3	1.5	0.3	4.0	2.1
CV[b]	26.3%	20.3%	26.3%	22.1%	17.3%	19.5%	4.2%	40.8%	23.6%
Skewness[a]	1.5	1.5	0.6	1.5	1.8	1.1	0.4	2.0	0.8
Range[a]	12.4	7.1	11.1	6.9	5.1	6.8	1.0	9.7	8.0
Min.[a]	4.5	4.5	4.6	5.6	6.1	5.2	6.8	7.1	6.0
Max.[a]	16.9	11.6	15.6	12.5	11.2	12.0	7.7	16.9	13.9
Count[c]	217	50	36	32	19	39	8	5	28

[a]All values expressed in l/100 km.
[b]Coefficient of variation.
[c]Represent number of trips per each driver.

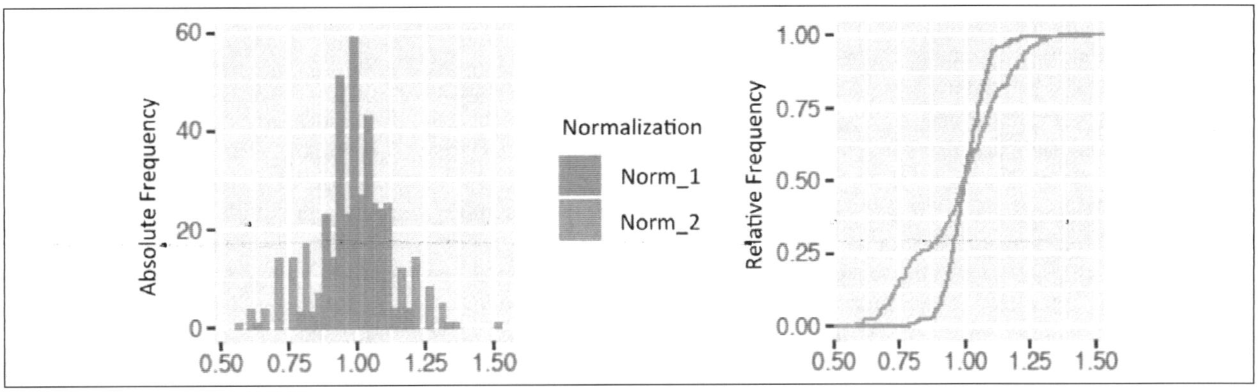

Figure 2. Empirical probability (chart on the left) and cumulative (chart on the right) distributions of normalized fuel consumption derived from the pool of vehicles subject to RDE tests in the period 2014–2016.

higher than the combined type-approval value of 5.5 l/100 km. Minimum fuel consumption within a trip was 4.5 l/100 km and maximum was 16.9 l/100 km. Only a few segments (five out of 217) had an average fuel consumption lower than the type-approval value of 5.5 l/100 km. The mean fuel consumption of all segments was 8.0 l/100 km (standard deviation = 2.1 l/100 km), 45% higher than the declared combined NEDC fuel consumption and higher than the average fuel consumption for complete RDE trips (7.3 l/100 km for all trips and 7.1 l/100 km for valid RDE trips). The coefficient of variation for the entire distribution of fuel consumption is 26.2%, which is significantly higher than that reported for the RDE trips (as expected because of a much higher variability of trip characteristics and number of drivers).

Data Set No. 2

The results of the analysis carried out on data set no. 2 are reported in Figure 2 and Table 6. The second normalization approach, as expected, shows significantly higher variability in the normalized CO_2 emissions. From a conceptual point of view, though, the second approach is the most accurate one, as in real life the average CO_2 emissions for a vehicle derive from all the trips that the vehicle has performed (which can involve totally different driving conditions).

Both Table 7 and Figure 2 show that the distribution of fuel consumption follows a fairly normal distribution. Given the way the normalization is carried out, both distributions are centered on 1. As expected, the first normalization has a smaller uncertainty and a peak of probability around 1. In both cases, the size of the fuel consumption uncertainty is significant. From the standard deviation it is possible to derive that around 95% of the observations are expected in a range of 35% of the average value in both directions. Considering that the RDE tests were carried out by a limited number of drivers and included only a limited number of routes in the same area, it can be expected that the actual uncertainty in fuel consumption of a single vehicle may be even higher. This is confirmed by the results obtained from data set no. 1, where the overall standard deviation is 26% of the average value.

Discussion

Given the intrinsic variability in the fuel consumption of a vehicle (as several factors have an effect on this variable), using the value measured during an RDE test (according to

Table 7. Main Statistics of the Empirical Distribution of Normalized Fuel Consumption

Normalization methods	Norm_1	Norm_2
Median	0.997	0.993
Mean	1	0.975
Min	0.639	0.544
Max	1.359	1.483
Standard deviation	0.079	0.178
Interquartile	0.1	0.27
Skewness	0.105	−0.164

the current protocol or a protocol specifically designed for the CO_2) as representative of the average fuel consumption of a vehicle on the road seems at least questionable. Data presented in this report (Table 5) show that the same vehicle driven for RDE purposes (though without any optimization objective) on a limited number of routes in the same area and with similar environmental conditions only accounts for a limited variability, and therefore the representativeness of the derived fuel consumption and CO_2 emissions would be bounded by the route selected for the test. In addition, as reported in the discussion of the results for the RDE tests carried out on the vehicle used for data set no. 1, the value obtained from a valid RDE test will probably be a conservative estimation of the average fuel consumption. Considering that vehicle manufacturers may even be able to select their own advantage-specific locations, environmental conditions, and driving circumstances from the many uncertainties intrinsic to an RDE test, the representativeness of the results of a valid RDE test could be further challenged.

The significant variability of the fuel consumption among different drivers (in the range 4.5–16.9 l/100 km for the vehicle used in the present study) shows that even when a more "central" estimation is achieved by means of a CO_2-RDE test (thus a test designed to achieve a representative real-world fuel consumption and CO_2 value, if possible to be defined), the benefits for users will be limited. Alternative approaches are available which, from a technical point of view, are simpler to use than attempting to develop a reliable and robust RDE type-approval test procedure for fuel consumption and CO_2.

Today's vehicles are all equipped with sensors and control units able to accurately estimate current fuel consumption, and most of them have already, or will soon have, connectivity features by which information on average fuel consumption over a certain period of time could be technically easy to collect and monitor [as an example in the U.S., and as of 2018 in Europe, all vehicles are equipped with a communication systems designed to inform the local authority about the occurrence of an accident (*24*)].

As in Europe, post-2020 CO_2 targets will be based on the CO_2 under WLTP testing conditions, a real-world fuel consumption monitoring system could be implemented and used to verify that any remaining gap between real-life and type-approval values does not significantly change over time. In this picture, the gap for a specific vehicle manufacturer (OEM) in the year t could be defined in the following way:

$$GAP(t) = \frac{\dfrac{\sum_{i=1}^{N} FC_i(t-1)\cdot D_i(t-1)}{\sum_{i=1}^{N} D_i(t-1)}}{\dfrac{1}{N}\sum_{i=1}^{N} FC_i^{TA}} \quad (1)$$

where $FC_i(t-1)$ is the specific fuel consumption of vehicle i in real life in the year $t-1$, $D_i(t-1)$ is the distance traveled by the vehicle i in the year $t-1$, FC_i^{TA} is the specific fuel consumption of vehicle i at TA, N is the number of vehicles registered by the OEM in the year $t-1$. Information concerning the evolution of this gap over time can be used, as an example, to correct fuel economy or CO_2 emission targets set for vehicle manufacturers by different governments around the world (by increasing or reducing them, depending whether the gap decreases or increases). This mechanism would ensure that the reduction of CO_2 emissions under type-approval conditions corresponds to a proportional time-invariant reduction in real life. In addition, a mechanism of this type would encourage the introduction of technologies which have an effect in real life, and would include those technologies whose effect is not, or is only partially, visible during the vehicle type-approval (without the need to implement and maintain additional complex mechanisms to award emission premiums for, for example, off-cycle technologies). Similar conclusions have also been reached by Greene et al. (*13*), in which the authors concluded that fuel economy needs to be rigorously and consistently measured over time to prevent the gap from widening.

Unfortunately a real-world CO_2 monitoring system cannot be used to provide more reliable a priori fuel consumption information to users. For this purpose, however, the idea itself of characterizing the fuel consumption of a vehicle with a single number is intrinsically inappropriate, given the stochastic nature of the variable (exemplified by the results collected in the present study). A possible alternative approach might be considered in which a user is instead provided with a simple modeling tool to calculate the fuel consumption of their vehicle in different conditions. By providing indications about driving preferences (for example, the usual route, the number of passengers and additional loads, the use of the on-board devices and instruments, the type of tires installed, etc.), a significant part of the uncertainty in fuel consumption can be reduced to a sufficiently low level. The same conclusions were reached by Greene et al. (*13*). The Green Driving tool developed by the JRC (*25*) can be considered as a first attempt in this direction. In the case that a link is established between the vehicle type-approval, the vehicle registration, and a tool such as Green

Driving, technical information on the vehicle and the results of the type-approval test could be used to characterize the vehicle efficiency in a way that could be sufficient to calculate real-world fuel consumption in a sufficiently reliable way. In the case of the Green Driving tool this was possible using the correlation tool CO_2MPAS to calculate CO_2 and fuel consumption under many possible conditions.

Conclusions

No matter how sophisticated and elaborate a test procedure is, inherently it cannot capture the multitude of factors affecting CO_2 emissions of cars in reality. WLTP is an extremely important tool for creating a level playing field for assessing the compliance of vehicles against set limits and for comparing the fuel consumption of individual vehicles but, ultimately, real savings in CO_2 emissions come during the real-world operation of cars. There are plenty of benefits to be harvested through initiatives that aim to reduce CO_2 emissions in the "real-world." New concepts need to be considered in addition to the traditional certification and monitoring schemes present in European legislation. New technologies may offer relatively simple and cost-effective tools for maximizing the results without necessarily putting additional burden in the certification process.

Building on the concept of real-world feedback and information, one could envisage a range of approaches. Drivers should be given more information for making purchase decisions that match their actual needs and everyday use of their vehicle, and should be empowered to provide visible feedback when such information is revealed as false or misleading. Vehicle labeling can be improved to offer such aspects with more detailed information. CO_2 monitoring in the real world should be considered, either by means of two-way voluntary interaction with the drivers or by establishing anonymized fuel consumption data collection instruments (for example, on-board fuel consumption data registration). Finally, in a world of applications and communication, the development of modeling tools and driver "communities," which could provide valuable feedback on vehicle performance and opportunities to reduce CO_2 emissions through efficient everyday practices, cannot be overlooked. In this sense, new policies should aim at an integrated framework covering all three aspects: official certification, real-world and lab-based emissions monitoring, and driver information-communication measures (as schematically reported in Figure 3). A basic tool aiming to provide a relevant platform to users and for establishing two-way communications between vehicle drivers and emission monitoring system has been embedded in the Joint Research Centre's Green Driving tool (*25*).

This three-pillar concept based on certification, monitoring, and user information can constitute a working bridge between reality and the laboratory, combining the benefits of all worlds and reducing the risks of future market or policy failures. It can provide a solid basis for sensible policy making in the decades to follow, enabling an accurate quantification of the CO_2 reductions achieved and revealing the great effort made by the industry to increase the energy efficiency of vehicles.

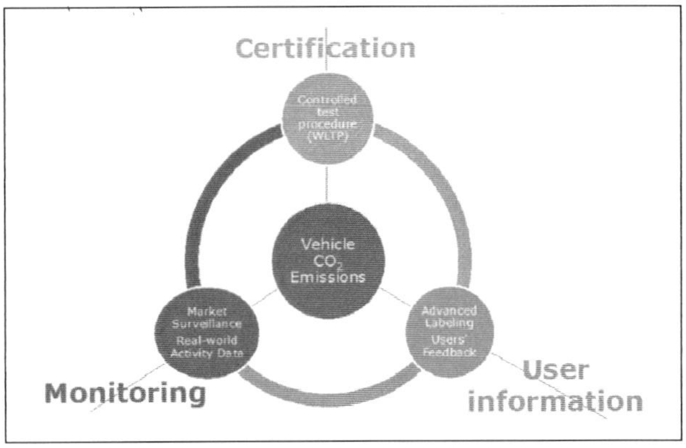

Figure 3. Schematic representation of the main elements that could form future integrated approaches for fuel consumption and CO2 emissions certification and labeling.

Acknowledgments

The activities described in the present paper have been carried out in the framework of the Administrative Arrangement n. 340201/2015/711841/SER/CLIMA between JRC and DG CLIMA. The authors are grateful to the staff of the Vehicle Emission Laboratory (VELA) of the JRC for carrying out all the RDE tests used in this analysis, and to all the colleagues from the Directorate General for Climate Action of the European Commission for the valuable comments provided.

References

1. Weiss, M., P. Bonnel, R. Hummel, U. Manfredi, R. Colombo, G. Lanape, P. Le Lijour, and M. Sculati.Analyzing On-Road Emissions of Light-Duty Vehicles with Portable Emission Measurement Systems (PEMS). *JRC Scientific and Technical Reports*. EUR 24697. 2011. Available at https://ec.europa.eu/clima/sites/clima/files/transport/vehicles/docs/2011_pems_jrc_62639_en.pdf. Accessed July 28, 2017.
2. Mellios, G., S. Hausberger, M. Keller, Z. Samaras, L. Ntziachristos, P. Dilara, and G. Fontaras.Parameterisation of Fuel Consumption and CO2 Emissions of Passenger Cars and Light Commercial Vehicles for Modelling Purposes. *JRC Scientific and Technical Reports*. EUR 24927. http://publications.jrc.ec.europa.eu/repository/bitstream/111111111/22474/1/co2_report_jrc_format_final2.pdf. Accessed July 28, 2017.
3. Fontaras, G., and P. Dilara. The Evolution of European Passenger Car Characteristics 2000–2010 and its Effects on Real-World CO2 Emissions and CO2 Reduction Policy. *Energy Policy* Vol. 49, 2012, pp 719–730.
4. Ligterink, N. E. Real World CO2 Emissions: Causes and Effects. Presented at the Low CVP conference, 2012.
5. Mock, P., U. Tietge, V. France, J. German, A. Bandivadekar, N. Ligterink, U. Lambrecht, J. Kulwhein, and I. Riemersma.From

Laboratory to Road: A 2014 Update of Official and 'Real-World' Fuel Consumption and CO2 Values for Passenger Cars in Europe. ICCT White paper. http://www.theicct.org/sites/default/files/publications/ICCT_LaboratoryToRoad_2014_Report_English.pdf. Accessed July 28, 2017.
6. Ligterink, N. E., and A. R. Eijik.A. Real-World Fuel Consumption of Passenger Cars. *TNO Report* n. 2016 R11258. doi:10.13140/2.1.1252.3527.
7. Reynaert, M., J. M. Sallee. Corrective Policy and Goodhart's Law: The Case of Carbon Emissions from Automobiles. Centre for economic policy research. http://cepr.org/sites/default/files/Reynaert%2C%20Mathias.pdf. Accessed July 27, 2017.
8. Tietge, U, et al. From Laboratory to Road - A 2015 Update of Official and 'Real-World' Fuel Consumption and CO2 Values for Passenger Cars in Europe. The International Council on Clean Transportation, 2015.
9. Zacharof, N., G. Fontaras, B. Ciuffo, S. Tasiakmakis, K. Anagnostopoulos, A. Marotta, J. Pavlovic, C. Codrea, and I. Hogdson. Review of In Use Factors Affecting the Fuel Consumption and CO2 Emissions of Passenger Cars. European Commission, 2016. JRC100150.
10. Fontaras, G., N.-G. Zacharof, and B. Ciuffo. Fuel Consumption and CO2 Emissions from Passenger Cars in Europe – Laboratory Versus Real-World Emissions. *Progress in Energy and Combustion Science*, Vol. 60, 2017, pp. 97–131.
11. Duarte, G. O., G. A. Gonçalves, and T. L. Farias. Analysis of Fuel Consumption and Pollutant Emissions of Regulated and Alternative Driving Cycles based on Real-World Measurements. *Transportation Research Part D: Transport and Environment*, Vol. 44, 2016, pp. 43–54.
12. Marotta, A., J. Pavlovic, B. Ciuffo, S. Serra, and G. Fontaras. Gaseous Emissions from Light-Duty Vehicles: Moving from NEDC to the New WLTP Test Procedure. *Environmental Science and Technology*, Vol. 49, No. 14, 2015, pp. 8315–8322.
13. Greene, D.L., A. J. Khattak, J. Liu, X. Wang, J. L. Hopson, and R. Goelts. What is the Evidence Concerning the Gap between On-Road and Environmental Protection Agency Fuel Economy Ratings? *Transport Policy*, Vol. 53, 2017, pp. 146–160.
14. Commission Implementing Regulation (EU) 2017/1152 of 2 June 2017 setting out a methodology for determining the correlation parameters necessary for reflecting the change in the regulatory test procedure with regard to light commercial vehicles and amending Implementing Regulation (EU) No 293/2012.
15. Commission Implementing Regulation (EU) 2017/1153 of 2 June 2017 setting out a methodology for determining the correlation parameters necessary for reflecting the change in the regulatory test procedure and amending Regulation (EU) No 1014/2010.
16. Ciuffo, B., and G. Fontaras. Models and Scientific Tools for Regulatory Purposes: The Case of CO_2 Emissions from Light Duty Vehicles in Europe. *Energy Policy*, Vol. 109, 2017, pp. 76–81.
17. Pathak, S., Y. R. Singh, V. Sood, S. K. Singal, and S. A. Channiwala. Investigation of Variation in Real World Fuel Economy of a Gasoline Car in Two Different Highway Traffic Conditions: An Indian Scenario. *SAE Technical Paper* 2014-01-1617. 2014, doi:10.4271/2014-01-1617.
18. Frey, H. C., K. Zhang, and N. M. Rouphail. Fuel Use and Emissions Comparisons for Alternative Routes, Time of Day, Road Grade, and Vehicles Based on In-Use Measurements. *Environmental Science and Technology*, Vol. 42, No. 7, 2008, pp. 2483–2489
19. Li, H., G. Andrews, B. Daham, M. Bell, et al., Impact of Traffic Conditions and Road Geometry on Real World Urban Emissions Using a SI Car. *SAE Technical Paper* 2007-01-0308, 2007, doi:10.4271/2007-01-0308.
20. Commission Regulation (EU) 2017/1154 of 7 June 2017 amending Regulation (EU) 2017/1151 supplementing Regulation (EC) No 715/2007 of the European Parliament and of the Council on type-approval of motor vehicles with respect to emissions from light passenger and commercial vehicles (Euro 5 and Euro 6) and on access to vehicle repair and maintenance information, amending Directive 2007/46/EC of the European Parliament and of the Council, Commission Regulation (EC) No 692/2008 and Commission Regulation (EU) No 1230/2012 and repealing Regulation (EC) No 692/2008 and Directive 2007/46/EC of the European Parliament and of the Council as regards real-driving emissions from light passenger and commercial vehicles (Euro 6).
21. Pavlovic, J., B. Ciuffo, G. Fontaras, and V. Valverde. An Insight into the Procedural Differences between NEDC and WLTP and their Possible Impact on CO2 Emissions from the Vehicle Type-Approval. Case of Vehicles with Internal Combustion Engines. Submitted for publication on *Energy policy*. Forthcoming.
22. Tsiakmakis, S., G. Fontaras, B. Ciuffo, and Z. Samaras. A Simulation-Based Methodology for Quantifying European Passenger Car Fleet CO2 Emissions. *Applied Energy*, Vol. 199, 2017, pp. 447–465.
23. Commission Regulation (EU) 2017/1151 of 1 June 2017 supplementing Regulation (EC) No 715/2007 of the European Parliament and of the Council on type-approval of motor vehicles with respect to emissions from light passenger and commercial vehicles (Euro 5 and Euro 6) and on access to vehicle repair and maintenance information, amending Directive 2007/46/EC of the European Parliament and of the Council, Commission Regulation (EC) No 692/2008 and Commission Regulation (EU) No 1230/2012 and repealing Commission Regulation (EC) No 692/2008.
24. European Commission. eCall in all new cars from April 2018. https://ec.europa.eu/digital-single-market/en/news/ecall-all-new-cars-april-2018. Accessed November 15, 2017.
25. Maineri, L., S. Tsiakmakis, V. Arcidiacono, G. Fontaras, B. Ciuffo, and C. Thiel.The JRC Green Driving Tool. Available at https://green-driving.jrc.ec.europa.eu/#/. Accessed July 27, 2017.

The Special Task Force on Climate Change and Energy (A0020T) peer-reviewed this paper (18-04408).

Article

TRR
JOURNAL OF THE TRANSPORTATION RESEARCH BOARD

Transportation Research Record
2018, Vol. 2672(2) 33–43

DOI: 10.1177/0361198118758393
journals.sagepub.com/home/trr
SAGE

Modeling the Impacts of Climatic Extremes on Interregional Freight-Transportation System

Arash Beheshtian[1,2], R. Richard Geddes[1], and Kieran P. Donaghy[2]

Abstract

The wellbeing of a nation is conditioned on an optimal distribution of commodity-flow. The freight transportation system, however, is exposed, and highly vulnerable, to climatic hazards. This becomes more problematic in the long term when climate change is projected to continue or worsen. Studying the long-term climate vulnerability of interregional commodity shipment is a challenging task given complexities inherent in (*a*) the dynamic formulation of factors such as location, production, the interaction of regional economies, and commodity distribution patterns, and (*b*) the stochastic arrival and intensity of natural hazards. This article studies the impact of projected climatic extremes in the State of New York on long-term interregional commodity-flow. It investigates the nature of and extent to which vulnerabilities in transportation infrastructure may cause sudden and profound changes in the routing assignment and cost of commodity-flows. We approximate commodity-flow assignment through a dynamic, long-term estimation of an interregional commodity-flow model reflecting the choices of households, market places, and firm locations and strategies. In the research here reported, we also utilized explicit treatment of trade in intermediate goods, the so-called *new economic geography* providing the behavioral foundation for production and interindustry and interregional trade, and an endogenous determination of capital investment and employment. The results show how and to what extent the (partial-)connectedness of New York State's freight-transportation network in time of disaster (*a*) causes an unmet demand for commodities across the nation's regional economies and (*b*) increases the shipment cost and effective-delivered price of commodity-flows.

The strong tie between the U.S. economy, national security, and the productivity of its interregional freight-transportation system is undeniable (*1*). The nation's freight-transportation system in 2015, for instance, distributed over $19 trillion worth of raw materials and finished goods, up from $17.7 trillion in 2012 (*2*). According to the 2015 Freight Analysis Framework (FAF[4]), by 2020, the annual freight shipment is projected to reach 24 billion tons, worth over $33.3 trillion, and maintaining an average annual growth rate of 3%.

Besides economic growth and development, interregional commodity-flow (IC-F) depends on the "availability" of supply-chain elements, in particular, physical elements of the transportation infrastructure. Freight-transportation system components, nevertheless, are under serious strain and confronted with potential vulnerabilities (*3*). The components are seriously challenged by: aging, congestion, overutilization, and underinvestment (*4*), losing gas-tax revenue to a new generation of fuel-efficient vehicles and alternative-fuel mobility (*5*), concerns over its contribution to 12% of the nation's CO_2 emissions (*6*), heavy politicization of its funding and planning (*7*), not proving highly adaptive to a massive technological revolution, as information and communications technologies are applied to optimize global supply chains and real-time routing systems (*8*), being capital intensive, with high maintenance or development costs, and being extremely vulnerable to man-made and natural hazards (*9*). Among the natural hazards, climatic extremes pose severe threats to physical infrastructure. Surface transportation infrastructure is vulnerable to sea-level rise (SLR), extreme temperatures, seasonal precipitation, and storm surges induced by hurricanes.

Despite the recent efforts toward "building resilience to climate", for example, U.S. Department of Transportation Climate Adaptation Plan 2014 (*10*), CMIP Climate Data Processing Tool (*11*), Vulnerability Assessment Scoring Tool (VAST) (*12*), Transportation Climate Change Sensitivity Matrix (*13*), FHWA drafts of Highways in the Coastal Environment: Assessing Extreme Events (*14*) and Highways in the River Environment-Floodplains, Extreme Events,

[1]Cornell Program in Infrastructure Policy, Cornell University, Ithaca, NY
[2]Department of City and Regional Planning, Cornell University, Ithaca, NY

Corresponding Author:
Address correspondence to Arash Beheshtian: ab2348@cornell.edu

Risk, and Resilience (*15*), the vulnerability of physical infrastructure to the projected long-term effects of climatic hazards has not been incorporated into guidelines.

In contrast to many studies investigating the vulnerability of physical assets (that is, property damage or "stock measures"), the purpose of this research is to assess the vulnerability of the IC-F in time ("flow measures") with respect to the trajectory of climatic extremes. For more on the distinction between these two measures, the reader may consult Rose (*16*), who argues that flow measures are "superior to stock measures." This research makes two key contributions to the literature. First, it models and solves dynamic IC-F for a real-size network and estimates the long-term evolution of the sectoral economy and freight-shipment pattern. Second, the proposed dynamic model is tested against the trajectory of natural hazards in New York State (NYS).

Literature Review

The impact of climate change on factors such as the transportation network, commodity supply chain, and (passenger-) travel behaviors is not the main focus of this research. For these topics, the reader may consult Beheshtian et al. (*17*), Koetse and Rietveld (*18*), and Halldórsson and Kovács (*19*). To study IC-F in a (post-)disaster scenario, we briefly review the following three modeling pillars.

Interregional Commodity-Flow under No-Disruption

Input–Output Models. Interregional commodity flows have been mostly studied through input–output (I-O) models (*20*). These studies are based on three pioneering models of "pure inter-regional input–output" by Isard (*21*), "multi-regional input–output" advanced by Leontief and Strout (*22*), and "column coefficient input–output" proposed by Moses (*23*).

I-O analysis fundamentally concerns the interactions of regional economies and industries. The I-O models consist of linear equations, where each equation represents the total production of a given commodity/sector as a function of (*a*) final demand of the same commodity/sector, (*b*) marginal, or technical, input coefficients assuming the level of interactions between different sectors, and (*c*) total production of commodities/sectors. The I-O models, however, have evolved dramatically through past decades. Extensions have been developed to incorporate details of economic activity and address concerns in energy industries (*24*), capture the effects of income generation for various household groups (*25*), and incorporate the activities of environmental pollution and elimination associated with economic activities (*26*).

Computable General Equilibrium (CGE) Models. Partial equilibrium approaches assume constant demand to estimate economic impacts from the supply side. This means that total impacts are estimated in the absence of outcomes from changes in demand (*27*). Intending to evaluate economic impacts from both the supply and demand sides, the CGE models (*a*) utilize price adjustments and analysis of behavioral response and (*b*) reflect non-linear damage functions and economies of scale. While Gillespie et al. (*28*) showed I-O analysis often overestimates economy-wide impacts, Rose and Liao (*29*) argue CGE models usually underestimate these impacts.

Static Versus Dynamic Modeling of IC-F. The sectoral interactions of economies, and hence IC-F models, are dynamic in nature. The IC-F in a given time period is a function of investment, production, and past marketing processes. The time variation between action and investment, as well as the existing temporal structure of interregional commodity production/shipment, have been less studied, and thus policymakers have not had the opportunity to study and test alternative policy assumptions by employing "comparative dynamics."

Friesz et al. (*30*) introduced a disequilibrium adjustment mechanism featuring the treatment of a general network topology and non-separable cost, supply, and demand functions. Ziliaskopoulos and Wardell (*31*) studied the importance of delays in commodity-flow assignment by proposing dynamic arc travel times and switching delays. They utilized a minimum cost path algorithm to solve the sub-problems. To understand the behavior of network decision makers, Nagurney et al. (*32*) modeled the presence of competition in a multi-level network of logistics, information, and finance. They implemented a discrete-time adjustment to describe the evolution of commodity shipments and prices to the equilibrium solution.

Agrawal and Ziliaskopoulos (*33*) developed an iterative formulation to model the market equilibrium and captured the shipper–carrier mechanism of the freight industry. Based on three components:(*a*) multimodal freight network simulation, (*b*) multimodal freight assignment, and (*c*) a multiple product intermodal shortest path procedure, Mahmassani et al. (*34*) developed a dynamic freight network simulation assignment platform. Donaghy (*35*) introduced two extensions to interregional freight shipment models: a partial-equilibrium static formulation and a dynamic formulation of location, production, and interaction.

Interregional Commodity-Flow under Climate Variability

IC-F under stress or climatic extremes could be studied in two distinct directions. The impact of natural disasters on flow itself, or the impact of disasters on flow as a derivate of another function known as "business interruption and infrastructure availability in time of disaster."

Table 1. Economic Regions and Centroids

Region	State(s)	BEA region[a]	Centroid
1	IL, MI, WI	Great Lakes	Chicago, IL
2	IN, OH	Great Lakes	Columbus, OH
3	PA	Mideast	Harrisburg, PA
4	NJ	Mideast	New Brunswick, NJ
5	NY	Mideast	Albany, NY
6	CT	New England	Hartford, CT
7	MA, RI	New England	Worcester, MA
8	ME, NH, VT	New England	Portland, ME
9	Rest of the U.S. (RUS)	–	Denver, CO[b]

[a]This demarcation on a national scale of economic regions represents rational location of productive forces and is in accordance with the Bureau of Economic Analysis (BEA) regionalizing systems.
[b]Denver, CO, is assigned as the centroid of RUS. Denver, which is one of the logistics hub locations, known as 'logistic cities,' hosts two major railways carriers, and is sited on the Interstate highway system.

Flow-focused models are cited in transportation and systems engineering literature. The reader should consult Chen and Miller-Hooks (*36*) for a description of its modeling approach and reviews. The latter, interruption-based models, however, could be investigated by examining direct damages, that is, "damages on stocks, which include physical and human capitals" (*37*), and by examining indirect damages, that is, "the system-wide impact of flow losses through inter-industry relationships" (*38*), also known as "higher-order effects" (*16*).

Okuyama (*37*) has also categorized economic modeling frameworks into (*a*) I-O models that are simple, though with inherent weaknesses, such as linearity and a lack of response to price changes. Hallegatte (*38*), Okuyama (*39*), and Rose and Wei (*40*) developed different applications of the I-O model; (*b*) CGE models, which are nonlinear, price-change responsive, and successfully reflect supply constraints. Rose and Liao (*29*), Tsuchiya et al. (*41*), and Hallegatte (*42*) are examples of CGE applications in disaster-assessment studies. Also, the dynamic stochastic general equilibrium model (DSGE) developed by Smets and Wouters (*43*) utilizes Bayesian generalized production functions to the analysis of macroeconomic monetary shocks; (*c*) the social accounting matrix (SAM) to examine the indirect effects on socio-economic agents, activities, and factors. Okuyama and Sahin (*44*) could be an example; (*d*) econometric models, which are capable of measuring long-run economic impacts. Coffman and Noy (*45*) is an example of the use of econometric models in a long-term disaster impact analysis.

Long-Term Interregional Commodity-Flow under Climate Variability Projections

The projection of interregional commodity flow in the face of climate variabilities faces two modeling challenges. First, dynamic analysis of IC-F models is still in its early stages. Second, projections for hazard-prone areas, disaster intensities, and intervals carry many uncertainties and are complex to model. Despite this, the proposed research simplifies this problem set by making some assumptions and introducing a methodology to predict the long-term impact of changing climate on IC-F.

Case Study and Assumptions

Economic Regions

The case study area covers the 48 adjoining-states of the continental USA. These states are divided into nine regions, as shown in Table 1. Each of the regions, called a freight analysis zone (FAZ), is represented by a centroid. The centroids are assumed to be the supply/market nodes being connected by transportation links. Therefore, all incoming and outgoing freight-traffic flow is attracted to and leaves the centroids, respectively.

Industrial Sectors

This research, akin to Ham et al. (*46*), assumes 13 sectors of the economy: agriculture, mining, construction, food production, chemical production, primary metals, fabricated metals, industrial machinery, electrical equipment, transportation equipment, other non-durable manufacturing, other durable manufacturing, and transportation, communication, utility (TCU), services, and government enterprises. It also assumes each sector is associated with one type of commodity, hence "sector" and "commodity" may be used interchangeably.

Freight-Transportation Infrastructures

Among the modes of freight distribution recognized by the U.S. Department of Commerce (USDOC) (i.e., truck, rail,

Figure 1. Top: Regional zoning. Middle: Infrastructure of roads. (Source: FHWA, Office of Operations.) Bottom: The US Freight Rail Network, primary corridors. (Source: The Research and Innovative Technology Administration [RITA], U.S. DOT.)

water, air, multiple modes and mail, pipeline, and others), this research only considers roads and railways, which together account for more than three-quarters of the total IC-F.

Roads. Without losing generality, this research considers the Primary Highway Freight System (PHFS), as well as the remaining National System of Interstate and Defense Highways ("the Interstate") that is not on the PHFS, as road infrastructure. The road infrastructure consists of transportation links connecting FAZs/centroids. While the centroids could be connected through the shortest path, if the shortest path is congested, it would be necessary to include link(s) providing each FAZ pair with an alternative path, assuring that commodity-flow switches from the shortest to a longer, but less congested, path. To this end, and aside from the nine centroids defined earlier, we introduce 26 transshipment nodes (i.e., nodes on which goods neither enter nor leave, but shipments may be assembled or divided). This increases the number of transportation links to 84.

The annual average daily traffic (AADT), also known as background traffic, was extracted from the 2011 Highway Performance Monitoring System (HPMS) released by the Office of Highway Policy Information, FHWA. The link-specific peak capacities estimated by the FAF[4] was used for road capacity.

Railways. Shipping heavy, low-value commodities, the U.S. commodity-rail network carries 3.7% of shipments measured by value, 12% of the weight, and nearly 18% of ton-miles (*47*). This infrastructure consists of the five major carriers: BNSF, Union Pacific (UP), CSX Transportation (CSXT), Norfolk Southern (NS), and Kansas City Southern (KCS) railway companies. Data for these carriers and for railway freight was extracted from Railroad Performance Measures, FHWA. The economy's regionalization and transportation infrastructure are shown in Figure 1.

Climatic Extremes

This research studied the impact of extreme events in NYS on long-term shipment of the IC-F. To do so, three types of climatic hazards to which NYS is most vulnerable were considered: SLR, 100-, and 500-year flood (YF). While these hazards are existing threats in NYS, long-term projections (based on representative concentration pathways, RCPs) suggest they are expected to become more intensive and frequent (*48*).

The trajectory of SLR will increase the frequency and intensity of hydro-meteorological hazards. In some parts of NYS, by the 2080s, the annual chance of today's 100-YF will be four times higher, and the corresponding flood surge height may reach up to 13 ft, according to the New York City Panel on Climate-Change (NPCC) (*49*). Reflecting the long-term climate vulnerability of the infrastructure of NYS, a series of 34 continuous sets was defined. Each set is tailored for a particular year starting in 2017, and includes three possible events: (*a*) SLR with no sudden event, (*b*) a 100-YF given SLR, and (*c*) a 500-YF given SLR. To simulate the current (2017) NYS events, the Flood Insurance Rate Map (FIRM), provided by the National Flood Insurance Program (NFIP), and Federal Emergency Management Agency (FEMA), was used. For the latter two events, the plan of 2020 and 2050 flood-prone areas developed by the NPCC was adopted. While there is no data for years 2018 to 2019 and 2021 to 2049, the flood surge in these years was simulated by projecting linear incrementation of inundation height between 2017, 2020, and 2050. Figure 2 shows the long-term vulnerability of the roads in a part of NYS.

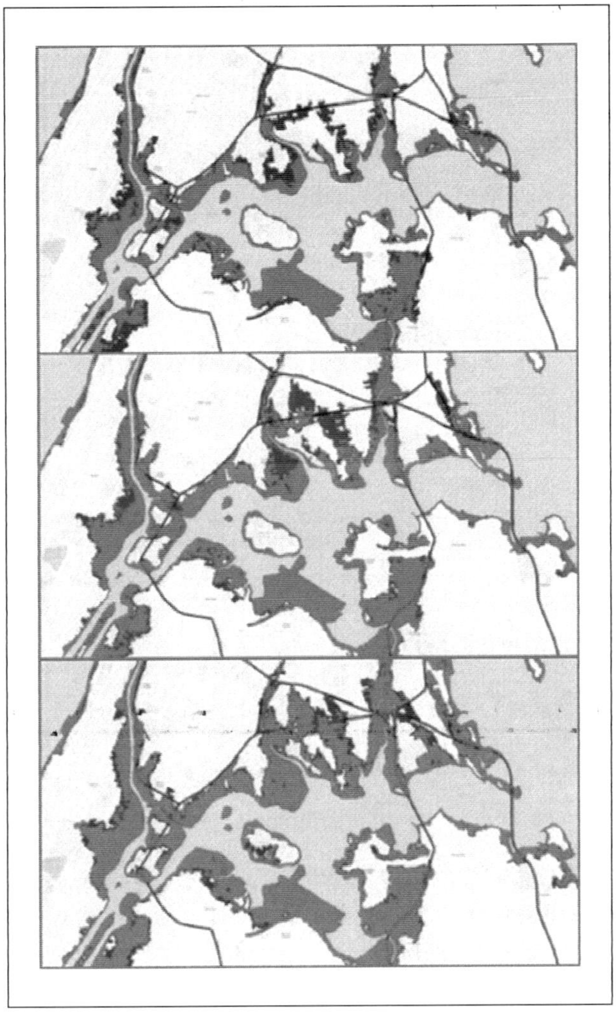

Figure 2. Top: Flood-prone areas to 100- and 500-YF in current year. (Source: FIRM, last update: May 2016.) Middle: Flood-prone areas to 100- and 500-YF in 2020s (projected). (Source: NPCC, 2015.) Bottom: Flood-prone areas to 100- and 500-YF in 2050s (projected). *(49)*.

Methodology

We adopt the dynamic IC-F model developed by Donaghy (*35*) and introduce some extensions to reflect the IC-F impacts of NYS's climate-vulnerable infrastructure, discussed earlier.

Modeling the Dynamic Interregional Commodity-Flow

The proposed model features a nonlinear mathematical program seeking the least cost associated with the production and distribution of thirteen types of commodities on two transportation modes, across 81 pairs of FAZs, and for a continuous 34 years starting in 2017. The objective function 1 shows the summation of (*a*) the production cost of h^{iv}_{lmrt} units of physical flow type *i*, in location *l*, to be shipped to and consumed in location *m*, with delivered unit-price of p^{iv}_{lmrt}, and (*b*) distribution cost associated with shipment of h^{iv}_{lmrt} between FAZ *l* and FAZ *m*, on route *r*, and by transportation mode *v*. the shipment and distribution costs are aggregated over time *t*.

$$\sum_t e^{-\Phi t}[\sum_{lmrv} D^v_{lmrt}(\sum_i h^{iv}_{lmrt} p^{iv}_{lmrt})] \tag{1}$$

Equation 2 reflects the delivered prices, p^{iv}_{lmrt}, defined as a function of (*a*) output unit-price (P^i_{lt}) of commodity type *i*, produced in location *l*, at time *t*, (*b*) modal shipping cost per ton-mile (p^v), (*c*) shipment distance (D^v_{lmrt}) on route *r*, mode *v*, between production plant *l* to market *m*, and (*d*) value to weight ratio (p^v).

$$p^{iv}_{lmrt} = P^i_{lt} + p^v \frac{D^v_{lmrt}}{q^i_x} \tag{2}$$

Equations 3–5 represent classic formulas of flow aggregation on roads linking assigned traffic on arc a' to its corresponding route r'.

$$D^{RD}_{lmr't} = \sum_{a'} d^{RD}_{a't} \delta^{a'RD}_{lmr'} \tag{3}$$

$$d^{RD}_{a't} = {}^{0}_{RD}d_{a't}\left[1 + 0.15\left(\frac{f^{RD}_{a't}}{k^{RD}_{a'}}\right)^4\right] \tag{4}$$

$$f^{RD}_{a't} = \sum_i \sum_{lmr'} h^{iRD}_{lmr'} \delta^{a'RD}_{lmr'} + TT^{RD}_{a't} \tag{5}$$

The flow assignment on railway tracks is adopted from Hwang and Ouyang (*50*). Equation 6 associates the flow assigned on route r'' and flow on track a'' which is a part of route r''. Equation 7 describes the congestion function on track a''. Equation 8 sums up the flows assigned on each direction of the track a''.

$$D^{RL}_{lmr''t} = \sum_{a''} d^{RL}_{a''t} \delta^{a''RL}_{lmr''} \tag{6}$$

$$d^{RL}_{a''t} = {}^{0}_{RL}d_{a''t} + \left(\frac{\alpha_{a''} l_{a''}}{100}\right) e^{\beta_{a''} f^{RL}_{a''t}} \tag{7}$$

$$f^{RL}_{a''t} = TT^{RL}_{a''t} + \sum_{ilmr''} h^{iRL}_{lmr''} \delta^{a''RL}_{lmr''} + \sum_{imlr''} h^{iiRL}_{mlr''} \delta^{a''RL}_{mlr''} \tag{8}$$

Equation 9, the flow conservation constraint, reconciles physical and value flows. Where $\wp^{iv}$, the modal share denotes the probability that good *i* will be shipped via mode *v* and, following Ham et al. (*46*), is taken to be a random utility function of average distance (AD) and average value of shipment of good i by available modes v. The modal shares of different commodity types are assumed; f_l^i reflects conversion factor adopted from Ham et al. (46).

$$\sum_r h_{lmrt}^{iv} = \sum_{r'} h_{mlr'}^{iRD} + \sum_{r''} h_{mlr''}^{iRL} =$$

$$\wp^{iv} \sum_j x_{lmt}^{ij} / \Phi_l^i q_x^i + \wp^{iv} \overline{FD}_{lmt}^{i} / q_x^i \qquad \forall i,\ v,\ l,\ m,\ t \qquad (9)$$

The part of industry i's product at location l that is shipped to all sources of final demand in just location m ($\overline{FD}_{lmt}^i$) can be approximated by Equation 10 in which $\overline{FD}_{lt}^i$ is the aggregate amount of the output from industry i at location l to be distributed to location m (as well as other locations) according to location m's relative share of total industrial output. In Equation 11, the final demand descriptor, D_{lmrt}^{*-1} is the inverse of the distance of the shortest actual route between locations l and m.

$$\overline{FD}_{lmt}^i = \overline{FD}_{lt}^i \frac{\sum_j X_{mt}^j}{\sum_m \sum_j X_{mt}^j} D_{lmrt}^{*-1} \qquad (10)$$

$$\overline{FD}_{lt}^i = X_{mt}^j / \Phi_l^i - \sum_m \sum_j x_{lmt}^{ij} / \Phi_l^i \qquad \forall i,\ l,\ t \qquad (11)$$

Equation 12, material balance constraint, ensures that shipment from industry i in location l, (X_{lt}^i), is equal to production by the industry in that location.

$$X_{lt}^i = \sum_m \sum_j x_{lmt}^{ij} + \sum_m \sum_k FD_{lmt}^{ik} \qquad (12)$$

The input demand specification is shown in Equation 13. Where ρ_m^j is the substitution parameter, κ_m^j is the economies-of-scale parameter for industry j in location m, ε_{lm}^{ij} is a share parameter, α_m^{ij} is a factor intensity parameter, π_m^j is a markup, P_m^j is the commodity type j's output price in location m, and p_{lm}^i will be a weighted average of the delivered prices of the input arriving via different modes v and routes r.

Equation 14 specifies the input aggregator function, c_{mt}^{ij}, where γ_m^{ij} is the scale parameter for the aggregator of inputs i in the production of good j.

$$x_{lmt}^{ij} = \varepsilon_{lm}^{ij} \kappa_m^j \alpha_m^{ij} / \pi_m^j \frac{P_m^j}{p_{lm}^i} (X_{mt}^j)^{\frac{\rho_m^j + \kappa_m^j}{\kappa_m^j}} (c_{mt}^{ij})^{-\rho_m^j} \qquad (13)$$

$$c_{mt}^{ij} = \gamma_m^{ij} \prod_l x_{lmt}^{ij\ \varepsilon_{lm}^{ij}} \qquad (14)$$

In Equation 15, p_{lm}^i denotes the effective delivered price (EDP) for all quantities of good i delivered from l to m by all routes.

$$p_{lm}^i = \sum_{rv} \frac{h_{lmr}^{iv} p_{lmr}^{iv}}{\sum_{rv} h_{lmr}^{iv}} \qquad (15)$$

Equation 16 reflects the dynamics between two continuous time-periods. Where υ_{lm}^{ij} is the adjustment parameter

$$\ln\left(x_{lmt+1}^{ij}\right) = \upsilon_{\text{lm}}^{ij} \ln\left(x_{lmt}^{ij}\right) - \left(\upsilon_{lm}^{ij} - 1\right) \ln\left(x_{lmt}^{ij}\right) \qquad (16)$$

These are subject to the nonnegativity and feasibility constraints 17–19.

$$f_{at}^v \geq 0 \qquad \forall v, a,\ t \qquad (17)$$

$$h_{lmrt}^{iv} \geq 0 \qquad \forall i,\ v,\ l,\ m,\ r,\ t \qquad (18)$$

$$x_{lmt}^{ij} > 0 \qquad \forall i, j,\ l,\ m,\ t \qquad (19)$$

Metrics of Inoperability

To study the impact of climatic hazards on IC-F, we define the following metrics:

a. Supply deficit (SD) (i.e., the ratio of unmet demand to the demand) in time of disaster. This metric (SD_{mt}^{iv}) is defined as the SD of commodity type i, in market m, shipped on all routes of transportation mode v, and from all supply nodes, in time t. This is equal to $\frac{\sum_{lr} h_{lmrt}^{iv} - \sum_{lr} h'^{iv}_{lmrt}}{\sum_{lr} h_{lmrt}^{iv}}$; where h'^{iv}_{lmrt} is the post-disaster physical flow.
b. The AD, in miles, every ton of commodity type i travels between production and market locations. The metric of average distance (AD_{mt}) is equal to $\frac{\sum_{lriv} h'^{iv}_{lmrt}}{\sum_{lrv} D'^{v}_{lmrt}}$; where D'^{v}_{lmrt} represents the distance between supply node l to market location m, at time of disaster in year t, and on route r of the transportation mode v.
c. The average effective delivered price (EDP_{mt}) in market location m, in year t, related to a ton of commodity of any type, shipped on transportation mode v from any of the origins. This is equal to $\frac{\sum_{lriv} p'^{iv}_{lmrt}}{\sum_{lriv} h'^{iv}_{lmrt}}$; where p'^{iv}_{lmrt} is the EDP for all quantity of good i, delivered from l to m by route r of the shipment mode v, at time t.The inoperability metrics discussed above are assumed for a yearlong timeframe.

Table 2. Description of Scenarios

Scenario	Number/type of annual climatic events		Annual likelihood of	
	100-YF $*PA_{100-YF} = 10^{-2}$ $**PD_{100-YF} = 3.8\times10^{-2}$	500-YF $PA_{500-YF} = 2 \times 10^{-3}$ $PD_{500-YF} = 9.6\times10^{-2}$	Scenario occurrence PA_{Sc}	Network disruption PD_{Sc}
1	1	–	10^{-2}	0.0383
2	–	1	2×10^{-3}	0.0958
3	1	1	2×10^{-5}	0.1342
4	2	–	10^{-4}	0.0767
5	–	2	4×10^{-6}	0.1917
6	2	2	4×10^{-10}	0.2684
7	2	1	2×10^{-7}	0.1726
8	1	2	4×10^{-8}	0.2301
9	3	–	10^{-6}	0.1150
10	4	–	10^{-8}	0.1534
11	5	–	10^{-10}	0.1917
12	–	3	8×10^{-9}	0.2876
13	–	–	0.98787	–

$*PA_{ce}$ reflects the annual arrival-likelihood of climatic event ce.
$**PD_{ce}$ reflects the likelihood/ratio of disrupted days, if event ce occurs.

Results and Discussion

Experiment Development

To understand the impact of climate change-induced hazards on IC-F, we developed five experiments. In each, we tested the proposed model against different environmental conditions.

Experiment I: IC-F under No Climatic Hazard. The first experiment represents the baseline model in which we simulated the long-term evolution of market, production, and IC-F assuming the physical elements of the supply chain experience no climatic extreme (i.e., the state of business as usual, or BAU). The dynamic model was tested for the years between 2017 and 2050.

Experiment II: IC-F under a 100-YF. The second experiment simulates the IC-F under a 100-YF. To this end, we adopted the long-term supply (i.e., production figures) and demand (i.e., market figures) from the BAU experiment, and simulated freight routing assignment for every year, separately, given the inoperability of the transportation physical infrastructure in time of disaster.

The supply and demand in flood-prone areas are assumed to be independent of the extremes, and equal to the BAU rates. While one may question this assumption, blocking ingoing and outgoing flows to and from the flooded FAZs, which we assumed earlier, is analogous to relaxing the production or demand values. Simulating the inoperability of transportation elements under a 100-YF, this study incapacitated the roads and tracks entering and leaving the FAZ's centroid.

Experiment III: IC-F under a 500-YF. Experiment III follows the same method used in Experiment II, yet simulates the impact of a 500-YF.

Experiment IV: IC-F under a Range of Scenarios. Through the first three experiments, the IC-F was modeled against one event. In the course of a year, however, the network experiences a range of events with different arrival likelihoods. For instance, the IC-F may experience one 100-YF per annum, and operate under the BAU state for the rest of the year. While the annual arrival chance of a 100-YF, for instance, is 1%, the IC-F is exposed to disruption on more than 1% of days. This means the event arrival likelihood is 1%, yet the aftermath lasts for weeks or months, as witnessed following superstorm Sandy. To reflect this, we assume a 100-YF, on average, will cause two weeks of disruption (i.e., 3.8% of a year). The disruption duration assumed following a 500-YF is five weeks, equal to 9.6% of a year.

Table 2 lists all possible scenarios, 13 in total, with more than 10^{-10} likelihood of occurrence that the IC-F may face as well as the disruption duration assumed in the aftermath of each scenario.

With respect to these scenarios, the inoperability metrics developed earlier are re-defined in Equations 20–22.

$$SD^{iv}_{mt\,Sc} = PD_{Sc} \frac{\sum_{lr} h^{iv}_{lmrt} - \sum_{lr} h'^{iv}_{lmrt}}{\sum_{lr} h^{iv}_{lmrt}} \quad (20)$$

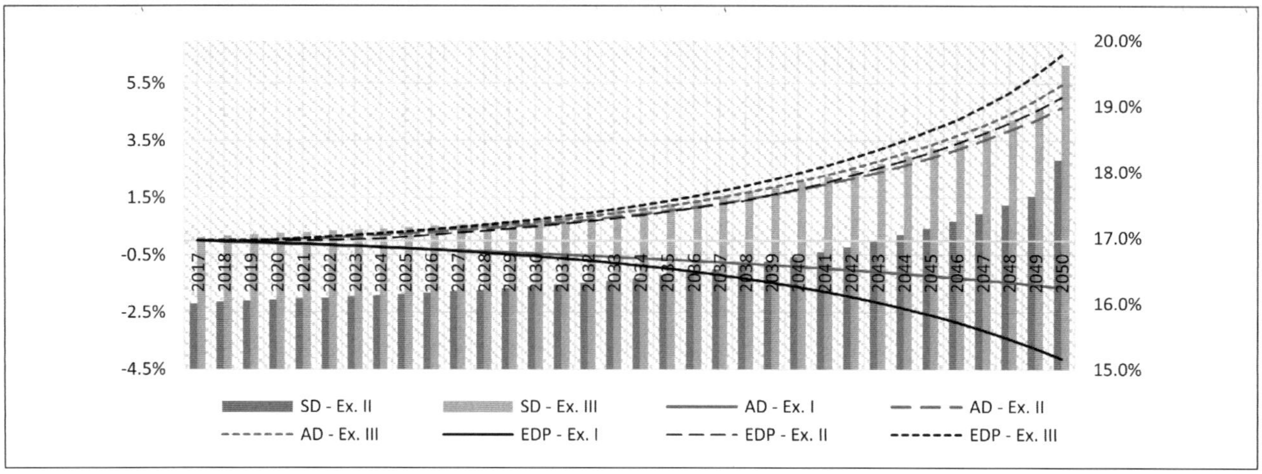

Figure 3. Results of Experiments I, II, and III.
Note: Supply-deficit values referred to the right axis and the rest referred to the left axis. The average distance (AD) is based on 2015 value: 1,683 mile/ton. The average effective delivered price (EDP) is based on present (2017) dollar value.

$$AD_{mt\,Sc} = PD_{Sc}\frac{\sum_{lriv} h'^{iv}_{lmrt}}{\sum_{lrv} D'^{v}_{lmrt}} + (1 - PD_{Sc})\frac{\sum_{lriv} h^{iv}_{lmrt}}{\sum_{lrv} D^{v}_{lmrt}} \quad (21)$$

$$EDP_{mt\,Sc} = PD_{Sc}\frac{\sum_{lriv} p'^{iv}_{lmrt}}{\sum_{lriv} h'^{iv}_{lmrt}} + (1 - PD_{Sc})\frac{\sum_{lriv} p^{iv}_{lmrt}}{\sum_{lriv} h^{iv}_{lmrt}} \quad (22)$$

Experiment V: Modeling the IC-F against All Possible Scenarios. While Experiment IV investigated the IC-F pattern against separate possible scenarios, this experiment investigated the IC-F in the face of all scenarios simultaneously. To this end, the inoperability metrics are defined as follows:

$$SD^{iv}_{mt} = \sum_{Sc=1}^{13} PA_{Sc} SD^{iv}_{mt\,Sc} \quad (23)$$

$$AD_{mt} = \sum_{Sc=1}^{13} PA_{Sc} AD_{mt\,Sc} \quad (24)$$

$$EDP_{mt} = \sum_{Sc=1}^{13} PA_{Sc} EDP_{mt\,Sc} \quad (25)$$

Each of the metrics is equal to the summation of scenario-specific metrics estimated in Experiment IV, weighted over the occurrence probability corresponding with scenarios (Table 2).

Modeling Output

First, we study how commodity-flows distribute during each of the possible events (i.e., BAU, 100-YF, and 500-YF). To this end, we compare the inoperability metrics under the first three experiments (results are shown in Figure 3).

There is no SD under the BAU experiment (Ex. I), since Equation 9, flow conservation constraint, assures no unmet demand across the market locations. In time of disaster, nevertheless, the vulnerable elements of the freight-transportation system may face flooding. This results in significant SD across the regions. The expected SD under Ex. II ranges between 16% and 18.2% for 2017 to 2050. The SD under 500-YF (i.e., Ex. III) and during the same time period is expected to be 17.1% to 19.6%. The SD in time of disaster and over 34 years grows by nearly 11%, while the IC-F tonnage is expected to increase by 41.8% by 2050 (i.e., 24.4 trillion tons in comparison with 18.1 trillion tons in 2017). This reflects the fact that while the flood-prone areas are projected to expand significantly, the number of vulnerable elements will not increase at the same rate. Most of the flood-vulnerable elements of the NYS transportation infrastructure in 2050 are also sited in 2017 flood-prone zones.

Figure 3 also compares the AD of every ton of commodity hauled between origin and market. While under Ex. 1 AD is projected to decrease by 1.65% in 2050, in time of disaster and through Ex. II and Ex. III, the AD—of delivered cargo—is projected to rise by 4.6% and 5.4%, respectively. This relates to disconnectedness of those shortest paths which have elements sited in flood-prone areas. The average EDP is conditioned on the length of the shortest yet open route between FAZ pairs (as shown in Equation 15). This urges the EDP metric to follow the AD's long-term pattern. The EDP through the Ex. II and Ex. III increases by 5.01% and 6.5%, respectively. This metric is projected to drop by 4.1% under a no-disruption scenario in 2050.

Next, through Ex. IV and Ex. V, the IC-F's long-term distribution pattern is studied during the scenarios embedding a range of events. The results of these experiments are compared in Figure 4.

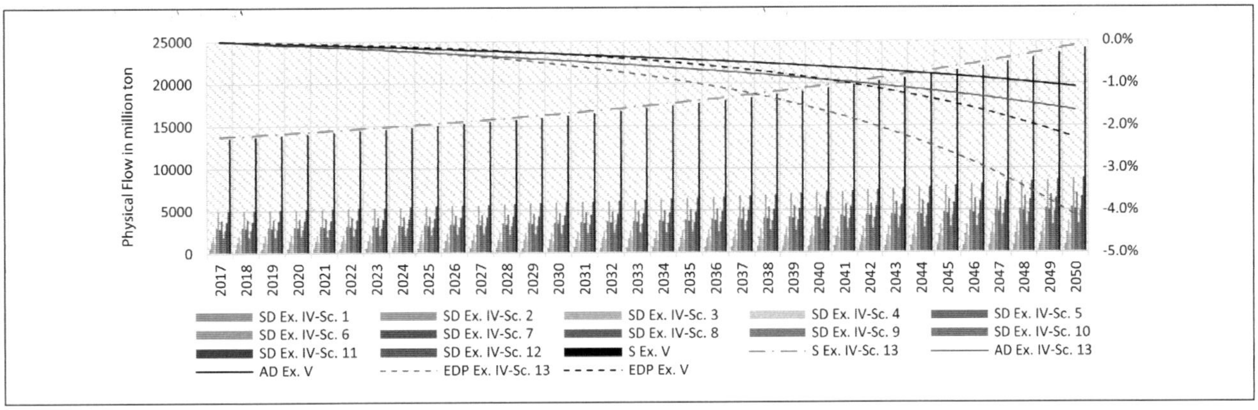

Figure 4. Results of Experiments IV and V.
Note: Bars (SD) and S values referred to the left axis. The rest referred to the right axis.

The first property of IC-F to be analyzed is SD averaged over FAZ pairs and all types of commodities. Through the scenarios (Sc.) 1 to 12 of Ex. IV, the values of SD increase in time. This relates to growing vulnerability of infrastructure elements to hazards. For example, through Sc. 8, Ex. IV, it is assumed (Table 2) that in a given year the IC-F experiences three types of events (i.e., no extreme event or BAU, one 100-YF, and two 500-YF). Assuming different disruption/recovery periods following each of the extreme events, the NYS's vulnerable elements will be incapacitated for 23% of the year. This results in 3.915 and 6.961 trillion tons of unmet demand for years 2017 and 2050, respectively. Under the last scenario of the Ex. IV, Sc. 13, there would be no SD, as IC-F faces no disruptive event.

Furthermore, the total physical flow through the fifth experiment is compared with the same metric during Sc. 13, Ex. IV. By drawing a comparison, under Ex. V, which is a combination of all scenarios included in the Ex. IV weighted average over the scenarios' likelihood, the freight-transportation system in year 2050 (when the IC-F tonnage and network vulnerability are at maximum levels), for instance, experiences only 282 million tons of SD, equal to 1.15% of the total tonnage of commodity projected for 2050 (i.e., 24.4 trillion tons).

The AD and EDP metrics under BAU state and Ex. V are also compared in Figure 4. The results show there is no considerable difference between the values of these metrics under Ex. IV, Sc. 13 (i.e., BAU state) and Ex. V. The AD and EPD values under no extreme event and by year 2050 drop to 1.64% and 4.13%, respectively. However, these metrics under Ex. V are expected to drop to 1.09% and 2.29%. While the NYS's vulnerable transportation elements experience dysfunctionality in 3.8% to 29% of the year—in the aftermath of Sc. 1 to 12 (Table 2), the AD and EDP metrics under the Ex. V are not substantial. This reflects the low arrival chance of disruptive scenarios (Sc. 1–12) which carry a cumulative annual likelihood of 1.213%.

Conclusion

If left unabated, the expected continued deterioration of the freight transportation system in the US would result in declining reliability and increased costs, and would put the nation at a disadvantage in the global economy. Investing in such capital-intensive, long-lasting infrastructure requires a forward-thinking strategic investment plan to maximize the long-term return on investment. The objective of potential investment in the freight-transportation system must be tailored to neutralize or otherwise minimize the burden of (in)direct associated costs on the public, and should be in line with the optimum performance of transportation infrastructures.

While a range of inoperability measures could be considered, this study examines the IC-F in the face of climatic disruptions. In summary, we found that, while NYS's transportation infrastructure accounts for a small share of the national freight-transportation system, the vulnerability of the state's physical infrastructure to SLR and flooding events will be one of the IC-F bottlenecks in time of climatic hazards. Such vulnerability, as discussed, may result in nationwide SD by 16% to 19.6%, increase in AD of delivered cargo by 4.6% to 5.4%, and increase in EDP to 6.5%. The reasons are several: (*a*) NYS is the supplier/market of 6.2% of the commodity tonnage and 11% of the dollar value for the nation as a whole; (*b*) the New York metropolitan area also carries 14% of the nation's container imports and 12% of its exports. Therefore, any interruption in commodity-flow distribution will have a significant effect and will ripple across the national and global economy.

Although climatic extremes are sudden events, they have costly and long-lasting consequences on physical infrastructure. Given this, and with respect to the projected increase in the intensity and frequency of climatic hazards, the potential impact of threats to climate-vulnerable infrastructure on the national economy is profound. Optimizing the long-term evolution of market locations and industrial development with respect to natural hazards projections, as shown in this article, could therefore provide IC-F with a more resilient distribution pattern.

There are several recommendations to improve the model. First, the physical network of IC-F is spatially dispersed and highly exposed to diverse natural hazards. This research only studied a part of the potential threats. More research is needed to study the impact of climatic extremes in more climate-vulnerable areas (specifically, in the southeast of the USA and the state of Texas, where coastal infrastructure and metropolitan areas are vulnerable to the changing climate) and modeled for different types of threats. Also, more detail on commodity types and industry-specific long-term projections can advance the accuracy of the model. Finally, the proposed model could be a base model to develop resilience-enhancing asset allocation models in which resources are invested prior to hazards to maximize IC-F's resilience in the event of disaster.

Acknowledgments

Research reported in this paper has been made possible by grant RD-8342831 from the U.S. Environmental Protection Agency STAR program, "Regional Infrastructure/Air Quality Planning and Global Change."

References

1. National Research Council. *Key Transportation Indicators: Summary of a Workshop*. National Academies Press, Washington, D.C., 2002.
2. Federal Highway Administration. U.S. Department of Transportation. Freight Analysis Framework. http://www.ops.fhwa.dot.gov/freight/freight_analysis/faf/index.htm. Accessed July 25, 2017.
3. National Freight Strategic Plan, 2015. https://www.transportation.gov/sites/dot.gov/files/docs/DRAFT_NFSP_for_Public_Comment_508_10%2015%2015%20v1.pdf. Accessed July 25, 2017.
4. Reinhardt, W. The Role of Private Investment in Meeting US Transportation Infrastructure Needs. *Public Works Financing*, Vol. 260, 2011, pp. 1–31.
5. Gas Tax Rates and the Highway Trust Fund, 2008. https://www.fhwa.dot.gov/publications/publicroads/08nov/03.cfm. Accessed July 21, 2017.
6. Freight and Air Quality Handbook, 2017. https://ops.fhwa.dot.gov/publications/fhwahop10024/sect2.htm. Accessed July 28, 2017.
7. TIGER Discretionary Grants, 2017. https://www.transportation.gov/tiger. Accessed July 27, 2017.
8. National Freight Strategic Plan, 2015. https://www.transportation.gov/sites/dot.gov/files/docs/DRAFT_NFSP_for_Public_Comment_508_10%2015%2015%20v1.pdf. Accessed July 25, 2017.
9. Beheshtian, A., 2016. Planning Resilient Infrastructure.
10. U.S. Department of Transportation Climate Adaptation Plan. https://www.transportation.gov/sites/dot.dev/files/docs/DOT%20Adaptation%20Plan.pdf. Accessed July 1, 2017
11. U.S. Department of Transportation's Coupled Model Intercomparison Project (CMIP) Climate Data Processing Tool, 2017. https://toolkit.climate.gov/tool/cmip-climate-data-processing-tool. Accessed July 1, 2017.
12. Vulnerability Assessment Scoring Tool (VAST), 2017. https://toolkit.climate.gov/tool/vulnerability-assessment-scoring-tool-vast. Accessed July 1, 2017.
13. Transportation Climate Change Sensitivity Matrix, 2017. https://toolkit.climate.gov/tool/transportation-climate-change-sensitivity-matrix. Accessed July 1, 2017.
14. *Highways in the Coastal Environment: Assessing Extreme Events*. Publication No. FHWA-NHI-14-006, 2014.
15. *Highways in the River Environment Floodplains, Extreme Events, Risk, and Resilience*. Publication No. FHWA-HIF-16-018, 2016.
16. Rose, A. Economic Principles, Issues, and Research Priorities in Hazard Loss Estimation. In *Modeling Spatial and Economic Impacts of Disasters*. Springer, Berlin Heidelberg, 2004, pp. 13–36.
17. Beheshtian, A., K. P. Donaghy, R. R. Geddes, and O. M. Rouhani. Planning Resilient Motor-Fuel Supply Chain. *International Journal of Disaster Risk Reduction*, Vol. 24, 2017, pp. 312–325.
18. Koetse, M. J., and P. Rietveld. The Impact of Climate Change and Weather on Transport: An Overview of Empirical Findings. *Transportation Research Part D: Transport and Environment*, Vol. 14, No. 3, 2009, pp. 205–221.
19. Halldórsson, Á., and G. Kovács. The Sustainable Agenda and Energy Efficiency: Logistics Solutions and Supply Chains in Times of Climate Change. *International Journal of Physical Distribution & Logistics Management*, Vol. 40, No. ½, 2010, pp. 5–13.
20. Georgia Tech Research Corporation, Parsons, Brinckerhoff, Quade & Douglas, and A. Strauss-Wieder, Inc. *NCHRP Report 732: Methodologies to Estimate the Economic Impacts of Disruptions to the Goods Movement System*. Transportation Research Board, Washington, D.C., 2012.
21. Isard, W. Interregional and Regional Input-Output Analysis: A Model of a Space-Economy. *The Review of Economics and Statistics*, Vol. 33, No. 4, 1951, pp. 318–328.
22. Leontief, W., and A. Strout. Multiregional Input-Output Analysis. In *Structural Interdependence and Economic Development*. Palgrave Macmillan, UK, 1963, pp. 119–150.
23. Moses, L. N. The Stability of Interregional Trading Patterns and Input-Output Analysis. *The American Economic Review*, Vol. 45, No. 5, 1955, pp. 803–826.
24. Spreng, D. T. *Net-Energy Analysis and the Energy Requirements of Energy Systems*. Praeger, 1988.
25. Miller, R. E., and P. D. Blair. *Input-Output Analysis: Foundations and Extensions*. Cambridge University Press, Cambridge, 2009.
26. Wier, M. Sources of Changes in Emissions from Energy: A Structural Decomposition Analysis. *Economic Systems Research*, Vol. 10, No. 2, 1998, pp. 99–112.
27. Richardson, H. W., Q. Pan, J. Park, and J. E. Moore (eds.). *Regional Economic Impacts of Terrorist Attacks, Natural Disasters and Metropolitan Policies*. Springer, Switzerland, 2015.
28. Gillespie, G., P. G. McGregor, J. K. Swales, and Y. P. Yin. The Displacement and Multiplier Effects of Regional Selective Assistance: A Computable General Equilibrium Analysis. *Regional Studies*, Vol. 35, No. 2, 2001, pp. 125–139.
29. Rose, A., and S. Y. Liao. Modeling Regional Economic Resilience to Disasters: A Computable General Equilibrium

Analysis of Water Service Disruptions. *Journal of Regional Science*, Vol. 45, No. 1, 2005, pp. 75–112.
30. Friesz, T. L., Z. G. Suo, and D. H. Bernstein. A Dynamic Disequilibrium Interregional Commodity Flow Model. *Transportation Research Part B: Methodological*, Vol. 32, No. 7, 1998, pp. 467–483.
31. Ziliaskopoulos, A., and W. Wardell. An Intermodal Optimum Path Algorithm for Multimodal Networks with Dynamic Arc Travel Times and Switching Delays. *European Journal of Operational Research*, Vol. 125, No. 3, 2000, pp. 486–502.
32. Nagurney, A., K. Ke, J. Cruz, K. Hancock, and F. Southworth. Dynamics of Supply Chains: A Multilevel (Logistical–Informational–Financial) Network Perspective. *Environment and Planning B: Planning and Design*, Vol. 29, No. 6, 2002, pp. 795–818.
33. Agrawal, B., and A. Ziliaskopoulos. Shipper-Carrier Dynamic Freight Assignment Model Using a Variational Inequality Approach. *Transportation Research Record: Journal of the Transportation Research Board*, 2006. 1966: 60–70.
34. Mahmassani, H., K. Zhang, J. Dong, C. C. Lu, V. Arcot, and E. Miller-Hooks. Dynamic Network Simulation-Assignment Platform for Multiproduct Intermodal Freight Transportation Analysis. *Transportation Research Record: Journal of the Transportation Research Board*, 2007. 2032: 9–16.
35. Donaghy, K. P. Some Extensions to Interregional Commodity-Flow Models. In *Socioeconomic Environmental Policies and Evaluations in Regional Science*. Springer, Singapore, 2017, pp. 161–175.
36. Chen, L., and E. Miller-Hooks. Resilience: An Indicator of Recovery Capability in Intermodal Freight Transport. *Transportation Science*, Vol. 46, No. 1, 2012, pp. 109–123.
37. Okuyama, Y. *Critical Review of Methodologies on Disaster Impacts Estimation*. Background paper for EDRR report, 2008.
38. Hallegatte, S. An Adaptive Regional Input-Output Model and Its Application to the Assessment of the Economic Cost of Katrina. *Risk Analysis*, Vol. 28, No. 3, 2008, pp. 779–799.
39. Okuyama, Y. Globalization and Localization of Disaster Impacts: An Empirical Examination. *CEFifo Forum*, Vol. 11, No. 2, 2010, pp. 56–66.
40. Rose, A., and D. Wei. Estimating the Economic Consequences of a Port Shutdown: The Special Role of Resilience. *Economic Systems Research*, Vol. 25, No. 2, 2013, pp. 212–232.
41. Tsuchiya, S., H. Tatano, and N. Okada. Economic Loss Assessment Due to Railroad and Highway Disruptions. *Economic Systems Research*, Vol. 19, No. 2, 2007, pp. 147–162.
42. Hallegatte, S. Modeling the Role of Inventories and Heterogeneity in the Assessment of the Economic Costs of Natural Disasters. *Risk Analysis*, Vol. 34, No. 1, 2014, pp. 152–167.
43. Smets, F., and R. Wouters. *Forecasting with a Bayesian DSGE Model: An Application to the EURO Area*. European Central Bank, Frankfurt am Main, Germany, 2004.
44. Okuyama, Y., and S. Sahin. Impact Estimation of Disasters: A Global Aggregate for 1960 to 2007. *Policy ResearchWorking Paper 4963*, TheWorld Bank, 2009.
45. Coffman, M., and I. Noy. Hurricane Iniki: Measuring the Long-Term Economic Impact of a Natural Disaster Using Synthetic Control. *Environment and Development Economics*, Vol. 17, No. 2, 2011, pp. 187–205.
46. Ham, H., T. J. Kim, and D. Boyce. Assessment of Economic Impacts from Unexpected Events with an Interregional Commodity Flow and Multimodal Transportation Network Model. *Transportation Research Part A: Policy and Practice*, Vol. 39, No. 10, 2005, pp. 849–860.
47. Featured Visualization. *Freight Flows by Highway, Railroad, and Waterway*, 2012. https://www.rita.dot.gov/bts/programs/freight_transportation/html/rail.html. Accessed June 21, 2017.
48. Beheshtian, A., Donaghy, K.P., Gao, H.O., Safaie, S. and Geddes, R. Impacts and Implications of Climatic Extremes for Resilience Planning of Transportation Energy: A Case Study of New York City. *Journal of Cleaner Production,* Vol. 174, 2018, pp. 1299–1313.
49. Orton, P., S. Vinogradov, N. Georgas, A. Blumberg, N. Lin, V. Gornitz, C. Little, K. Jacob, and R. Horton. New York City Panel on Climate Change 2015 Report Chapter 4: Dynamic Coastal Flood Modeling. *Annals of the New York Academy of Sciences*, Vol. 1336, No. 1, 2015, pp. 56–66.
50. Hwang, T., and Y. Ouyang. Assignment of Freight Shipment Demand in Congested Rail Networks. *Transportation Research Record: Journal of the Transportation Research Board*, 2014. 2448: 37–44.

The Special Task Force on Climate Change and Energy (A0020T) peer-reviewed this paper (18-06515).

Article

TRR
JOURNAL OF THE TRANSPORTATION RESEARCH BOARD

Transportation Research Record
2018, Vol. 2672(2) 44–53

DOI: 10.1177/0361198118794043
journals.sagepub.com/home/trr
SAGE

Prioritization of Climate Change Adaptation Interventions in a Road Network combining Spatial Socio-Economic Data, Network Criticality Analysis, and Flood Risk Assessments

Xavier Espinet[1] and Julie Rozenberg[1]

Abstract
Climate change puts at risk all current and future transport projects. Investing proactively in climate adaptation of transport infrastructure is paramount to providing resilience and sustainable transport systems that may promote social and economic growth. Despite the importance of such investments, the financial resources of many road administrations are constrained, creating an urgent need to allocate these resources efficiently to capture the highest social, environmental, and economic benefits. This paper aims to tackle this issue by presenting a methodology to prioritize climate change adaptation interventions based on a set of economic, social, and risk reduction criteria. This method uses a network-wide approach to the road system in order to capture co-benefits, redundancies, and costs of disruption of road segments due to flood events. All underlying data in each of the criteria is part of a geospatial database that includes the location of agriculture, fishery production areas, high poverty, and flood maps. The methodology was developed at the request of the World Bank Africa Team to support the Government of Mozambique to prioritize climate change adaptation intervention in two central provinces, Zambezia and Nampula. After combining criteria for agriculture, fishery, poverty, network criticality, and hazard risk, the results suggest that most roads in the coastal districts of these provinces could be identified as top priority for climate change adaptation interventions.

Providing farmers with reliable connectivity is key for achieving sustainable growth (*1*). Similar to most rural areas of Sub-Saharan Africa, rural accessibility is very low in the Zambezia and Nampula provinces of Mozambique and most farmers practice subsistence agriculture (*2*). A recent study found that in these two regions, even if all existing roads were upgraded, connectivity would not exceed 40% (*3*).There is thus an urgent need for transport investments. Improving farmers' connectivity to markets would increase their revenues and reduce poverty, while also improving the national economy (*e.g.* by reducing food import needs) (*4*).

The benefits of transport investments also depend on the reliability of transport services under extreme and unexpected events (*5*). Mozambique is affected by river floods and storm surges that damage the transport infrastructure and create disruptions in the road network (*6*). Since the Mozambican road network has a low redundancy, those disruptions sometimes isolate communities for months.

Given a finite budget and more than 30,000 km of classified roads in the country (plus 30–50,000 km of unclassified roads), the question arises of how to prioritize interventions that will trigger the highest agricultural benefits and maintain the reliability of the network under extreme circumstances. The World Bank Africa team approached the authors to address this challenge and to develop a methodology to support the Government of Mozambique in prioritizing climate change adaptation interventions. Therefore, the methodology was

[1]Office of the Chief Economist for Sustainable Development, The World Bank, Washington, D.C.

Corresponding Author:
Address correspondence to Xavier Espinet: xespinetalegre@worldbank.org

developed based on the needs of the National Road Authority of Mozambique.

This paper seeks to address this challenge by identifying critical and vulnerable roads in Zambezia and Nampula provinces and then supporting the selection of interventions that can ensure the reliability of the road network for rural communities. Following a brief background, the paper is divided into two sections. The first defines the methodology used to identify districts where the road interventions should be prioritized. The methodology provides a detailed approach to analyze road criticality, based on different metrics of network performances. The second section presents the case study in the road networks in the Zambezia and Nampula provinces in northern Mozambique. The goal of this section is to select the top three districts of greatest priority in each province.

Background

Roads, bridges, and culverts are the most vulnerable elements of the transport system with regard to adverse and extreme weather conditions (*7–9*). Disruption of those elements may have severe cascading reactions along the entire transport system, isolating rural villages or towns and disrupting trade or business operations, among other effects (*10*). Climate change has exacerbated these vulnerabilities, increasing the frequency of disruptions along the transport systems, and has challenged every single step of the transport planning process. Transport infrastructure has become one of the sectors most vulnerable to climate change in the civil infrastructure realm (*11*, *12*).

In particular, the magnitude of future floods, storms, and extreme weather events fostered by climate change poses challenges to transport engineering design, maintenance, rehabilitation scheduling, and disaster preparedness plans. The current design standards, practices, and building codes have become obsolete, as they were developed based on historical weather patterns (*13*). Failing to update standards, practices, and codes for new and existing transport projects will lead to an increase of maintenance, repair, and rehabilitation budgets for almost every road administration agency around the world (*14–16*), which will exacerbate the enormous negative impacts of climate change on transport users (*17–20*).

In response to these recognized threats to transport infrastructure, many transport professionals and stakeholders in the transport field have started to move toward a more holistic and proactive mind set, incorporating consideration of climate change into transport planning, engineering and design (*21*, *22*). This approach would lower hazard risk, social impact, life-cycle cost, and infrastructure degradation rates, creating more sustainable transport asset management and overall increasing the resilience of the transport system to natural disasters (*9*, *11*, *23*).

As part of sustainable transport asset management, a network-level approach is required in order to capture fully the vulnerability of the transport system and in order to evaluate the real impact of the disruption (*24*, *25*). When a bridge is disrupted due to extreme floods, users will look for a second-best route to reach their original destination. If that second-best route leads to much longer travel time, or there is no second-best route available, the disruption would be very severe. This method, if applied systematically to every link of the network, would identify the most vulnerable segments and would test the overall resilience of the transport network (*26*, *27*).

This paper combines all these aspects and introduces an innovative road infrastructure prioritization framework that combines socio-economic considerations and climate change risk reduction based on a network systems approach.

Methodology

The goal of this section is to present a comprehensive methodology for prioritization of roads and districts based on geospatial data, network analysis, and risk assessment. The methodology assesses the exposure and the vulnerability of the road network to river floods and storm surges, and calculates the disruption risk in each district under multiple future scenarios. Finally, the criticality and risk are combined to identify which districts should be a priority for investment.

The prioritization is based on two pillars: (i) socio-economic criticality and (ii) current and potential hazard risk to the roads. At the same time, the socio-economic criticality is composed by five attributes: (a) network criticality, (b) proximity to potential agriculture cluster, (c) proximity to potential fishery cluster, (d) current agriculture production, and (e) poverty rate in adjacent districts. The second pillar, hazard risk, is calculated using flood maps for current climate and climate change scenarios, combined with vulnerability function of the different pieces of infrastructure in the road system, such as bridge, culvert, or road surface. The hazard risk is then calculated as expected annual damage to the infrastructure using 10 return periods and four climate scenarios (current climate and low, medium, and high levels of climate change). Figure 1 shows a flow chart of the methodological approach, the corresponding figure number (in brackets) to the pillar on the results section.

The following sections describe in detail each of the two pillars and its components, and presents the results for applying the methodology to the Zambezia and Nampula provinces.

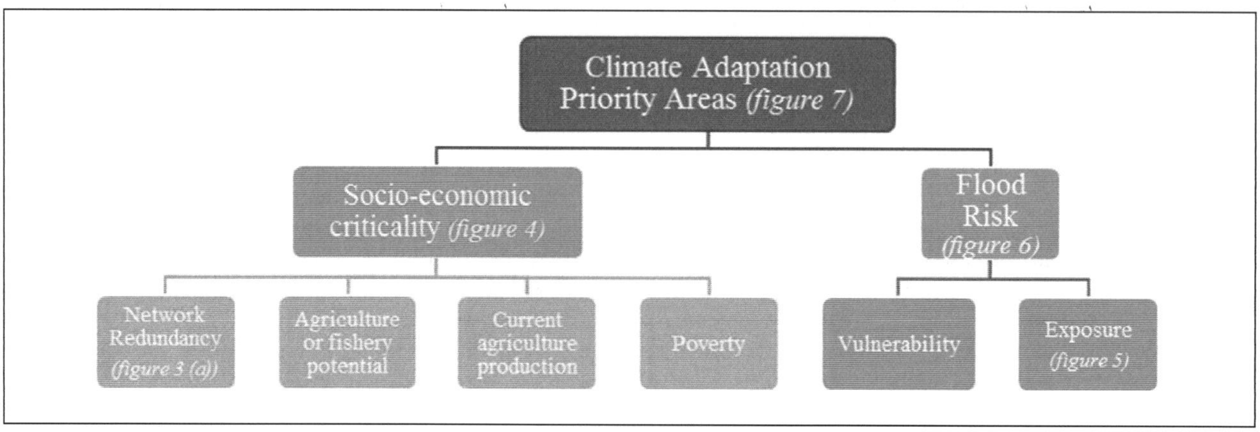

Figure 1. Flow chart with methodological approach.

Criticality

In this analysis, we define the criticality of a link as the loss of network performance if this link is removed from the network (*28*, *29*). The metrics for evaluating the performance are cumulated as road user cost (RUC), defined as U.S. dollars per vehicle ($/vehicle) and based on road condition (measured using the International Roughness Index [IRI]) and data from the HDM4 model.

We first calculate these three variables in a baseline with the complete existing network. We then remove each of the links one by one and we recalculate the performance of the network with these three indicators, based on the assumption that travelers will take the second-best route if the optimal route is disrupted. The optimal and second-best routes are the routes that minimize the total RUC between the origin and destination. The additional cost, time, and distance of travel due to the disruption of a link are aggregated over all origin–destination (OD) pairs and assigned to the disrupted link to determine its criticality.

Lack of redundancy is only one way of measuring the criticality of a link, and other attributes may be relevant for decision-makers. In the current study, proximity to poverty hotspots, current agricultural production, and fishery or agriculture potential are particularly important for determining where to intervene in the network. We thus create five categories of criticality for each of these attributes, combine them using equal weights, and produce a new criticality map. The assumption of equal weight between criticality attributes could be altered to answer certain priorities of the decision-maker. The equal weights approach was agreed with local stakeholders during two workshops held in Zambezia and Nampula in January of 2017. For instance, if the decision-maker aims to invest in roads to promote fishery, we could give a higher weight to that attribute.

The criticality values at road level are then aggregated at the district level. The new criticality index is now used to rank districts within one province. To aggregate criticality at the district level, we sum criticality indices weighted by the length of the roads. The Government of Mozambique wants to invest in feeder roads in these two districts – that is, roads feeding traffic to the main highway. The criticality of the primary road is thus less important for deciding where to invest, since there is only one main highway, and a separate investment program is maintaining the critical sections of this highway.

Exposure

The second step of the analysis is to identify the exposure of the network links to climate events. We focus our analysis on flood risk and we combine floods coming from riverine and storm surge. Fluvial flood hazard layers were derived from the SSBN Global Flood Model (GFM) developed by Sampson et al. (*30*). GFM is a large-scale hydrodynamic modeling structure designed to enable flood modeling in data scarce regions. The model itself is a fully 2D hydraulic model, using custom processed terrain data to provide flood hazard data at ~90 m resolution. It is driven by the global Regional Flood Frequency Analysis (RFFA) that has been derived from the gauged flow records of thousands of rivers across the world. The SSBN GFM was used to produce water depth maps for current climate, climate change, and land-use change conditions for 10 different flood recurrence intervals (from 5 to 1000-year return period).

Vulnerability

The next step of the analysis is to calculate the vulnerability of the road infrastructure to different water levels. This vulnerability is expressed as the cost of repairing or rebuilding bridges, culverts, and road surfaces when a flood occurs.

The bridges inventory used comes directly from the national road administration (ANE) GIS database and uses the conditions surveyed in 2015. There is actually no inventory for culverts, although we found it necessary to include culverts in the analysis as they can easily be damaged by flood and disrupt the road. In order to estimate the location of culverts the river map was intersected with the road network. The intersection points that were not included in the bridge inventory were assumed to be culverts. The estimation has been calibrated with data collected in the field for some roads. Our estimation seems to be very conservative and a larger number of culverts is more realistic.

For bridges and culverts the damage is modeled as a percentage of total replacement cost. The damage would be calculated as the ratio between the flood water level and the water level it was designed for. If the flood water level is lower than the water level it was designed for, no damage would be expected. However, it is assumed that the percentage of damage is proportional to the percentage of additional water level above the design. We add two factors, drainage capacity rate (Dc) and condition rate (Cr) to be applied to the damage. The Dc rate is set at 0.7 (as default), and it reduces the total amount of water that the structure can drain. The Cr is applied directly to the damage as a multiplying factor, to account for the fact that structures in poor condition will incur more damage than structures in excellent condition. The condition rates are set to 0.7, 0.5, and 0.3 for good, fair, and poor condition. The following simplistic damage equation is used:

$$D_i = \left(\frac{WLcc_i - WLd^*Dc}{WLd}\right)^* 1/Cr^*Rc \qquad (1)$$

- WLd = water level design comes from the design standard of 20 years for culverts and 100 years for bridges on primary and secondary roads (probability is based on past events)
- $WLcc_i$ = water level for climate change projection for return period i
- Dc = drainage capacity rate
- Cr = condition rate
- Rc = replacement cost

The damage cost on road surface is based on three thresholds; 0.2, 0.5, and 1.5 m of water above the surface. It is assumed that roads are built at ground level (no embankments). For each of the three water level thresholds we assigned a percentage of total replacement cost if surpassed, different for paved or unpaved: [1, 2, 20%] for paved and [20,50,100%] for unpaved.

Risk to the Infrastructure (Expected Annual Damage to the Infrastructure)

The final step is to combine the hazards, vulnerability, and exposure to estimate the flood risk that each district faces annually, expressed in terms of damage to the infrastructure.

Damages are calculated for 10 different flood events (5, 10, 20, 50, 75, 100, 200, 250, 500, 1000-year return period), and for four different climate scenarios (one based on past climate and three on climate change projections), in each of the river basins in the two provinces. Individual events are then aggregated, for each climate scenario, into expected annual losses (EAD). To calculate the EAD the trapezoidal rule was used, using the losses associated with each event and its probability of occurrence as the inverse of the return period:

$$EAD = \frac{1}{2}\sum_{i=1}^{n}\left(\frac{1}{T_i} - \frac{1}{T_{i+1}}\right)(D_i + D_{i+1}) \qquad (2)$$

where i is an integer between 1 and 10, T_i is the ith return period, and D_i is the damage to the infrastructure corresponding to T_i.

For the prioritization process we use only the risk expressed in terms of damage to the infrastructure and we do not calculate the risk for users. The risk in terms of user costs will be used for assessing the benefits of interventions. This is because the risk in terms of damage to the infrastructure can easily be calculated at the link level and then aggregated at the district level, while the risk to users has to be calculated at the country level – since travelers crossing the flooded area can come from all over the country – and is not relevant at the district level.

Prioritization

The last step of this section is to prioritize districts for future interventions. The choice to prioritize districts and not roads is a pragmatic one, to improve coordination and create economies of scale.

The prioritization is a combination of the criticality identified before and the risk to the infrastructure. A prioritization matrix is built to combine criticality and risk, giving a score of one to a very low risk road and five to a very high risk road. For example, a district identified as low priority will have a low risk, and will be low priority for investment, no matter the level of risk. At the other side of the matrix a road identified as highly critical will range from medium to very high priority and therefore will be a priority for investment, as any disruption will cause a huge burden for users and the local economy.

Results

Baseline Network Model

We first selected the nodes and roads of Mozambique's network that were useful for the analysis. Even if we focused on Nampula and Zambezia, we modeled the national network since trips from and to those provinces have origins and destinations in the entire country. ANE provided the classified road network, and we completed it with an assessment of the conditions of unclassified roads in the provinces of Nampula and Zambezia using an innovative smartphone-based technique during summer of 2016 (*31*, *32*). This last set of roads was merged into the classified roads to obtain a complete transport network. The clean new network is composed of 798 links containing paved and unpaved roads, and primary, secondary, tertiary and unclassified roads (Figure 2).

We identified, in consultation with local authorities, 138 major OD nodes, including major population clusters, seaports, airports, border crossings, and, of special interest for this study, agriculture production clusters in the provinces of Nampula and Zambezia. The agriculture clusters were identified based on the results of an agriculture production potential model, SPAM, developed by the International Food Policy Research Institute.

The road network contains information on the traffic volume which was provided by ANE and collected in the field for unclassified roads using annual average daily traffic (AADT). Using the measured traffic volumes and a simplified travel demand model, we estimated the traffic between each OD pair. A few roads, totaling 275 km, were left out of the analysis as they did not give access to any of the identified nodes.

Criticality

This criticality analysis ranks most sections of the National Highway 1 (NH1) as highly or very highly critical because secondary routes are in much worse condition and impose much higher costs on road users (Figure 3, left panel). Some secondary and tertiary roads in the districts of Gurue, Milange, and Namarroi, in the northwestern part of Zambezia, are also identified as highly critical; the density of roads in that area is low and therefore the disruption of some of those links will lead to a large detour representing a cost of more than $30/vehicle. Those districts are of special interest as they give access to one of the main agriculture production clusters.

When poverty, current agriculture, and agriculture or fishery potential are combined with the lack of redundancy in the network, some other districts are identified as critical (Figure 3*b*). Specifically, two districts in the south of Nampula—Angoche and Moma—are found to be highly critical due to both a very high poverty rate and proximity of fishery production. While poverty is more critical in Zambezia and in some parts in the south

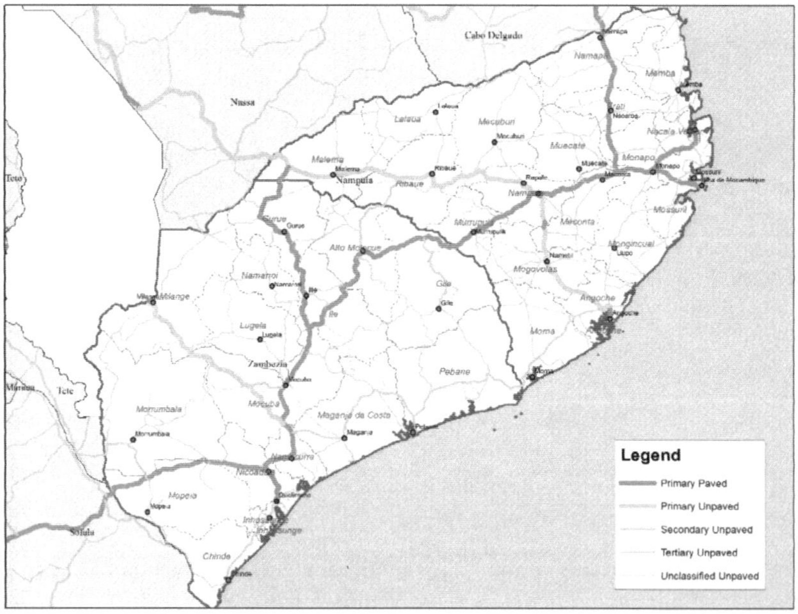

Figure 2. Road map of Nampula and Zambezia provinces.

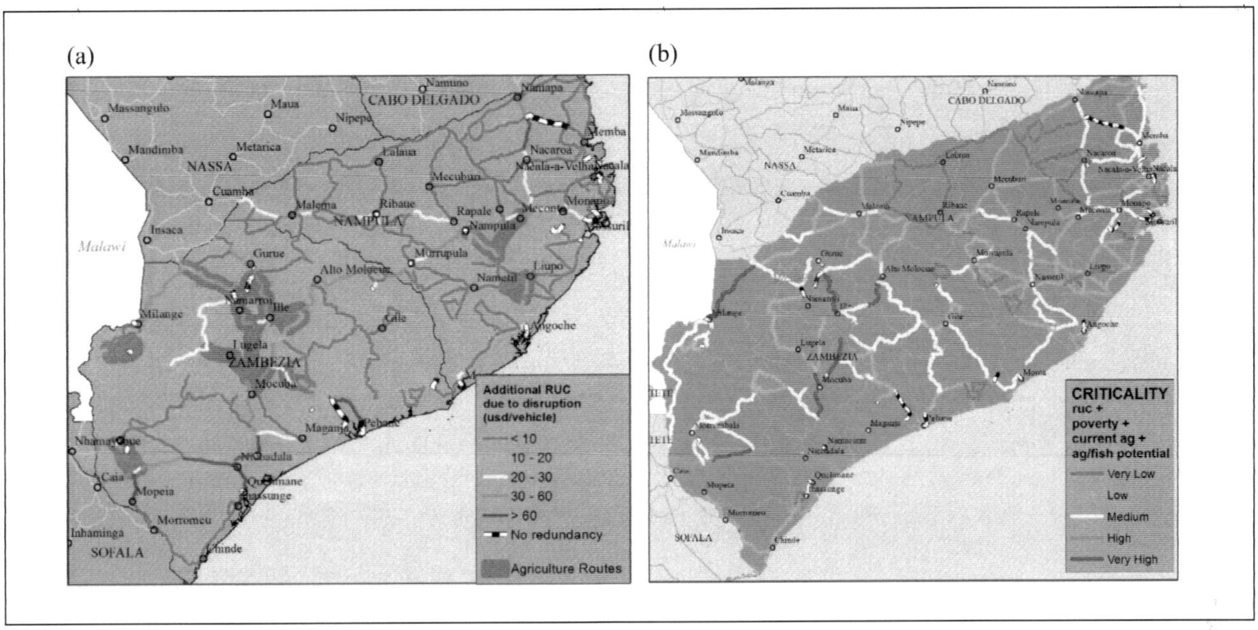

Figure 3. (a) Criticality map for Nampula and Zambezia based on RUC increase (in $/vehicle); (b) and criticality as aggregation of redundancy, poverty, current agriculture, and agriculture or fishery potential.

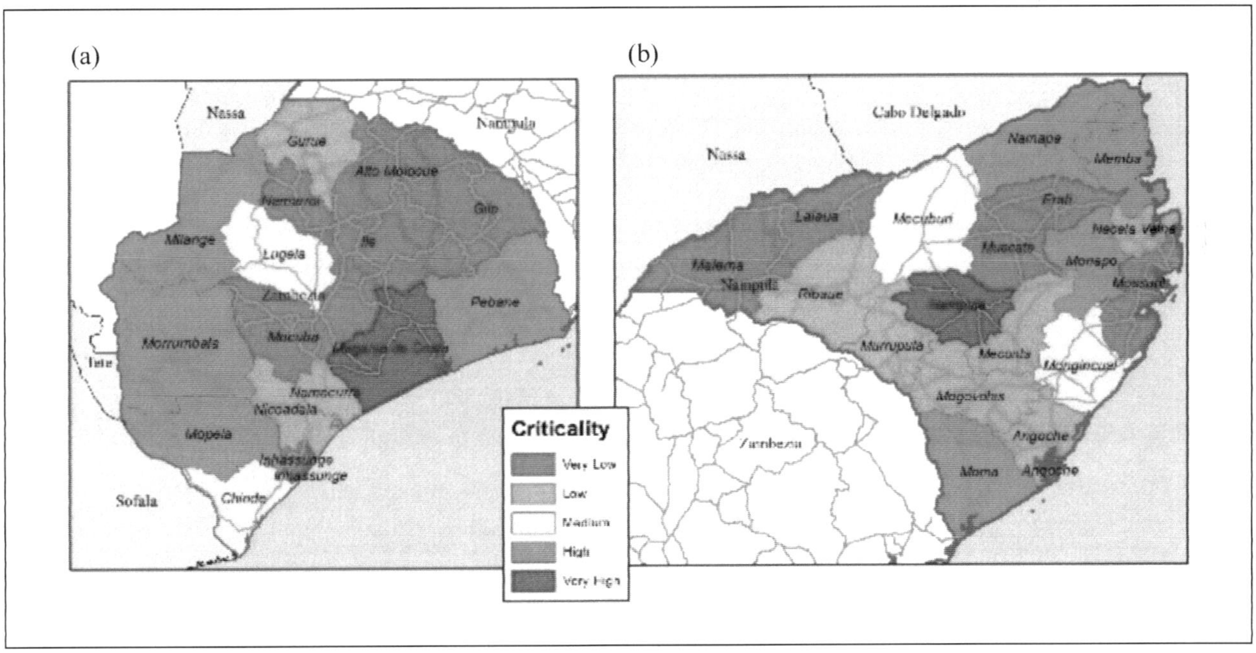

Figure 4. Criticality map at a district level without primary roads: (a) Zambezia; (b) Nampula.

of Nampula, when incorporating fishery, several roads appear to be critical in the coastal districts in Nampula. In Zambezia, a few more districts in the south appear to be critical, in particular, Maganja, Mocuba, and Mopeia.

For this reason, we created criticality maps in which the primary roads were excluded from the aggregation (Figure 4). In this case the most critical districts in Zambezia are Maganja, Pebane, and Morrumala. For

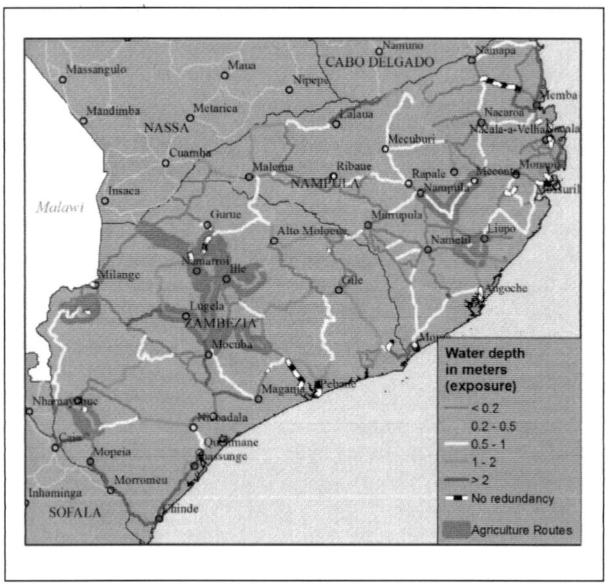

Figure 5. Exposure map for Nampula and Zambezia's road network based on maximum water depths.

Nampula, Nampula district, Memba, Namapa, Moma, and Monapo are the most critical districts.

Exposure

We overlay the road network with the different flood maps and represent in Figure 5 the water depth that each link is exposed to for a 1 in 100 flood event under past climate conditions. Roads around the Zambezi River around the border between Tete, Sofala and Zambezia provinces are highly exposed to 100-year flood events with water levels higher than 2 m. These high water levels affect some of the agriculture clusters in some parts of Zambezia province, particularly those in the Zambezi and Licungo River basins. On the other hand, most roads in Nampula have a medium or low exposure with maximum water level less than 1 m, except for a few roads in the northern part of Nampula in the Lurio River basin (Figure 5).

Vulnerability and Risk

The expected annual damage to the infrastructure was thus calculated for each road and summed again at the district level. Figure 6 shows the risk for the infrastructure in millions of U.S. dollars for each district, excluding losses from the primary roads (the same maps including the primary roads can be found in the annex). Following the exposure results, the most vulnerable districts in Nampula are Moma, Memba, and Namapa. The first is located in the delta of the Molocue River, and Memba and Namapa are in the Lurio River basin. In Zambezia, in addition to the districts in the Zambezia and Licungo River basins, the districts of Ile, Maganja, and Pebane are identified as highly vulnerable with more than £1 million of expected annual losses. Those three districts have numerous smaller tributary rivers and that, combined

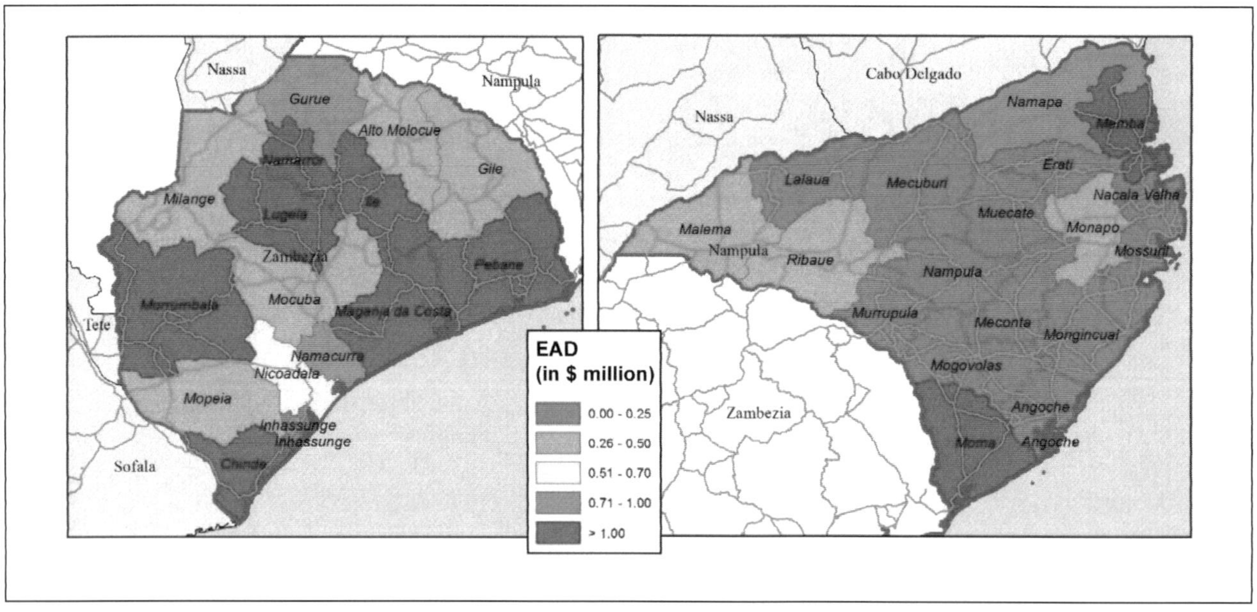

Figure 6. District risk expressed as expected annual damage to the infrastructure under current climate condition, excluding primary roads for Zambezia (left) and Nampula (right).

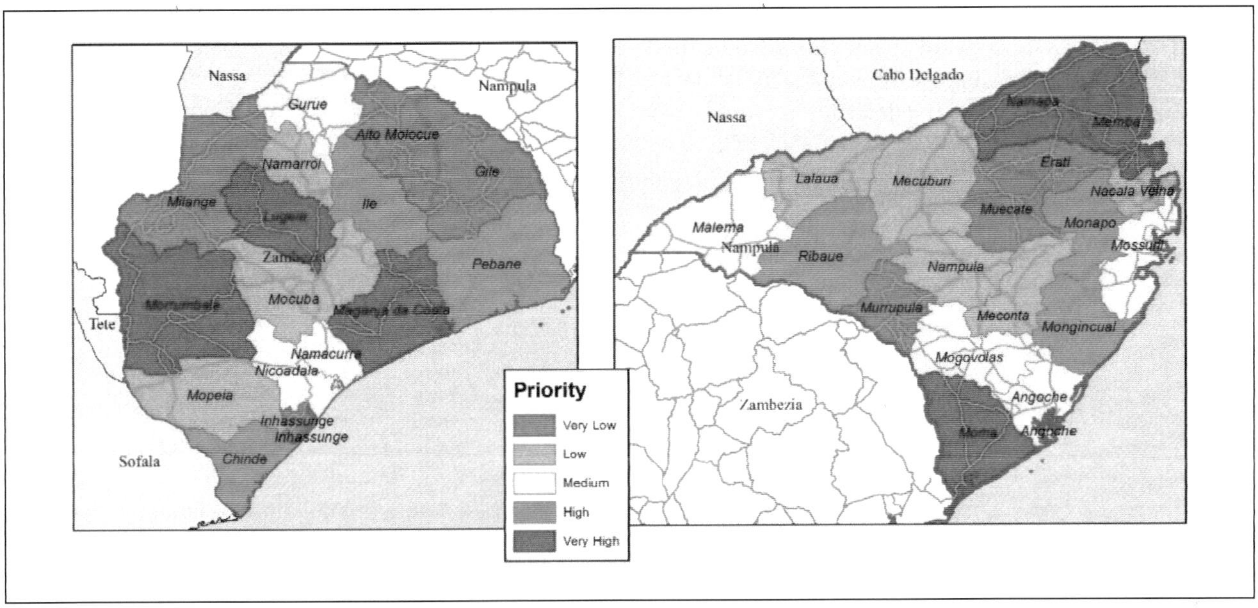

Figure 7. Risk map at district level excluding primary roads for Zambezia (left) and Nampula (right).

with the poor condition of the road network, makes those districts very vulnerable to flood events.

Prioritization

Figure 7 presents the prioritization maps produced using the risk to infrastructure (due to current climate) and the criticality (as a combination of redundancy, current agriculture, agriculture or fishery potential, and poverty). The results are driven by the individual road values and aggregated at the district level. These maps exclude values from primary roads. The top priority districts in Zambezia are Morrumbala and Maganja. The top priority districts in Nampula are Moma, Memba, and Namapa.

Conclusions

This paper presents several advances to the traditional prioritization analysis of transport investment. First, it presents a new method to quantify system-wide benefits due to improving the resilience of transport networks from disruption caused by hazard events. Using a network approach, this method quantifies the cost benefits to road users from upgrading second-best routes. This approach also quantifies the non-monetized benefits in terms of reduction of isolated trips and avoided disruption time. Second, it combines socio-economic criteria with vulnerability and flood risk reduction. We have demonstrated that by combining all of these criteria we can prioritize both sustainable and resilient interventions in the most critical districts.

Incorporating network benefits has proven to be very useful to identify intervention methods that may help to make a more resilient transport network, by improving overall network performance. Incorporating network-wide benefits is particularly useful to identify interventions that help to improve the condition of redundancy in case of disruption along main corridors. These interventions support the achievement of network resilience, as the interventions offer improved alternatives and second-best routes. In the traditional project-level economic analysis, such benefits would not have been considered.

Applying the methodology to the road networks in the Nampula and Zambezia provinces, it is difficult to extract broad conclusions as to where the climate change adaptation interventions should be prioritized. Each of the districts selected in the two provinces has a different combination of factors that triggers a "high" or "very high" prioritization ranking. In general, however, districts that have a high flood risk have been identified as high priority. This follows the overall objective of the project, to identify intervention approaches that would reduce the impact of climate change. Therefore, intervention in districts with high flood risk would lead to higher benefits from climate change adaptation strategies. Flood risk is particularly high in Zambezia, which contains many rivers and creeks; as such the added assessment of socio-economic criteria helped to identify the most critical district among several with high flood risk.

One of the lessons learned during the development of this study was that, early on, the team realized that the first road network data supplied by the Government of Mozambique had a large number of gaps. Especially in the rural areas of interest, many rural roads were not included in the map. However, those roads are critical for the analysis and they normally serve the most deprived and impoverish communities. We identified those gaps by using other data sources as Open Street Maps and satellite imaginary. Additionally, in order to perform a network redundancy analysis, the road network needs to be topologically correct (no gaps). In our case that required a large effort of data cleaning of the GIS road network. One of the critical assumptions in terms of flood risk analysis is to assume that the roads are built at ground level. The resolution of the digital elevation model used in the flood maps did not recognize road embankments or other flood management structures. It was not feasible to develop refined hydrological models that would capture those structures for the entirety of the two provinces. Therefore, the values of the risk assessment should be treated as indicative as we recognize that we may be overestimating real cost.

The authors are currently working on a follow-up study that includes the use of an innovative method to prioritize road interventions in a selected district. This method is embedded in techniques for decision-making under uncertainty (*29*), to deal with deep uncertainty in cost-benefit analyses. Such an approach identifies the combinations of variables that may cause investments to fail and will help decision-makers to select interventions that are robust to many future scenarios. Some of the preliminary findings indicate that repair and maintenance of bridges and culverts are the most cost-effective interventions in order to reduce the impacts of climate change. Those interventions will cost around £15 million per district but will result in a net present value over 20 years of more than £20 million in most cases.

Additionally, after this pilot project, other World Bank teams are starting to implement the methodology presented in this paper in order to support their government counterparts to prioritize climate change adaptation interventions. This includes rural roads in Albania, the national road network in Vietnam, and even in other types of transport network such as an inland waterway project in Bangladesh.

Acknowledgments

The authors acknowledge the assistance received from their counterparts in Mozambique; ANE (*Administração Nacional de Estradas*) and *Fundo de Estradas* which did a great job preparing the field data collection and provided all the necessary equipment and additional data. The authors are also grateful for the support, help, and suggestions received from other members of staff of the World Bank, especially Marianne Fay, Cecilia Briceno-Garmendia, Kulwinder Rao, Satoshi Ogita, Atsushi Imii, and Laura Bonzanigo. The authors are grateful for the support received through the Korean Green Growth Trust Fund.

Author Contributions

The authors confirm contribution to the paper as follows: study conception and design: Xavier Espinet, Julie Rozenberg, Laura Bonzanigo; data collection: Xavier Espinet; analysis and interpretation of results: Xavier Espinet, Julie Rozenberg; draft manuscript preparation: Xavier Espinet and Julie Rozenberg. All authors reviewed the results and approved the final version of the manuscript.

References

1. *Global Launch of IFAD Rural Development Report 2016.* https://www.ifad.org/ruraldevelopementreport. Accessed June 26, 2017.
2. Iimi, A., F. Ahmed, E. C. Anderson, A. S. Diehl, L. Maiyo, T. Peralta-Quirós, and K. S. Rao. *New Rural Access Index: Main Determinants and Correlation to Poverty.* Publication ID 2867657. Social Science Research Network, Rochester, NY, 2016.
3. Iimi, A., and K. S. Rao. *Mozambique: Spatial Analysis of Transport Connectivity and Growth Potential.* The World Bank, Washington, D.C., forthcoming.
4. Bell, C., and S. van Dillen. How Does India's Rural Roads Program Affect the Grassroots? Findings from a Survey in Upland Orissa. *Land Economics*, Vol. 90, No. 2, 2014, pp. 372–394. https://doi.org/10.3368/le.90.2.372.
5. Cervigni, R., A. M. Losos, P. Chinowsky, and J. L. Neumann. *Enhancing the Climate Resilience of Africa's Infrastructure: The Roads and Bridges Sector.* Publication 110137. The World Bank, 2016, pp. 1–0.
6. Building Resilience into Mozambique's Roads—Quickly. *The World Bank - News - Feature Story*, March 16, 2016. http://www.worldbank.org/en/news/feature/2016/03/16/building-resilience-into-mozambiques-roads-quickly
7. *Special Report 290: Potential Impacts of Climate Change on U.S. Transportation.* Transportation Research Board of the National Academies Washington, D.C., 2008.
8. Nemry, F., and H. Demirel. *Impacts of Climate Change on Transport. A Focus on Road and Rail Transport Infrastructures.* Publications Office of the European Union, Luxembourg, 2012.
9. European Environment Agency. *Adaptation of Transport to Climate Change in Europe.* Publication 8/2014. European Environment Agency, Publications Office of the European Union, Luxembourg, 2014.
10. Koetse, M. J., and P. Rietveld. The Impact of Climate Change and Weather on Transport: An Overview of Empirical Findings. *Transportation Research Part D: Transport and Environment*, Vol. 14, No. 3, 2009, pp. 205–221. https://doi.org/10.1016/j.trd.2008.12.004.
11. Meyer, M., A. Amekudzi, and J. O'Har. Transportation Asset Management Systems and Climate Change.

Transportation Research Record: Journal of the Transportation Research Board, 2010. 2160: 12–20.
12. Bollinger, L. A., C. W. J. Bogmans, E. J. L. Chappin, G. P. J. Dijkema, J. N. Huibregtse, N. Maas, T. Schenk, M. Snelder, P. van Thienen, S. de Wit, B. Wols, and L. A. Tavasszy. Climate Adaptation of Interconnected Infrastructures: A Framework for Supporting Governance. *Regional Environmental Change*, Vol. 14, No. 3, 2014, pp. 919–931. https://doi.org/10.1007/s10113-013-0428-4.
13. Lisø, K. R. Integrated Approach to Risk Management of Future Climate Change Impacts. *Building Research & Information*, Vol. 34, No. 1, 2006, pp. 1–10. https://doi.org/10.1080/09613210500356022.
14. Schweikert, A., P. Chinowsky, X. Espinet, and M. Tarbert. Climate Change and Infrastructure Impacts: Comparing the Impact on Roads in Ten Countries through 2100. *Procedia Engineering*, Vol. 78, 2014, pp. 306–316. https://doi.org/10.1016/j.proeng.2014.07.072.
15. Schweikert, A., X. Espinet, S. Goldstein, and P. Chinowsky. Resilience Versus Risk. *Transportation Research Record: Journal of the Transportation Research Board*, 2015. 2532: 13–20.
16. Espinet, X., A. Schweikert, N. van den Heever, and P. Chinowsky. Planning Resilient Roads for the Future Environment and Climate Change: Quantifying the Vulnerability of the Primary Transport Infrastructure System in Mexico. *Transport Policy*, Vol. 50, 2016, pp. 78–86.
17. Keener, V. W., J. J. Marra, M. L. Finucane, D. Spooner, and M. H. Smith (eds.). *Climate Change and Pacific Islands: Indicators and Impacts: Report for the 2012 Pacific Islands Regional Climate Assessment*. Island Press, Washington, D.C., 2012.
18. McBeath, J. Institutional Responses to Climate Change: The Case of the Alaska Transportation System. *Mitigation and Adaptation Strategies for Global Change*, Vol. 8, No. 1, 2003, pp. 3–28. https://doi.org/10.1023/A:1025840627213.
19. Hambly, D., J. Andrey, B. Mills, and C. Fletcher. Projected Implications of Climate Change for Road Safety in Greater Vancouver, Canada. *Climatic Change*, Vol. 116, No. 3–4, 2013, pp. 613–629. https://doi.org/10.1007/s10584-012-0499-0.
20. Bruin, K. de, R. B. Dellink, A. Ruijs, L. Bolwidt, A. van Buuren, J. Graveland, R. S. de Groot, P. J. Kuikman, S. Reinhard, R. P. Roetter, V. C. Tassone, A. Verhagen, and E. C. van Ierland. Adapting to Climate Change in the Netherlands: An Inventory of Climate Adaptation Options and Ranking of Alternatives. *Climatic Change*, Vol. 95, No. 1–2, 2009, pp. 23–45. https://doi.org/10.1007/s10584-009-9576-4.
21. Ebinger, J. O., and N. L. Vandycke. *Moving Toward Climate-Resilient Transport: The World Bank's Experience from Building Adaptation into Programs*. Publication 102406. The World Bank, Washington, D.C., 2015, pp. 1–62.
22. Bennett, C. R., I. D. Greenwood, T. F. P. Henning, and S. Tighe. Integrating Climate Change into Road Asset Management: *Technical Report 2017*. Publication 114641. The World Bank, Washington, D.C., 2017, pp. 1–76.
23. Meyer, M., and B. Weigel. Climate Change and Transportation Engineering: Preparing for a Sustainable Future. *Journal of Transportation Engineering*, Vol. 137, No. 6, 2011, pp. 393–403. https://doi.org/10.1061/(ASCE)TE.1943-5436.0000108.
24. Berdica, K. An Introduction to Road Vulnerability: What Has Been Done, Is Done and Should Be Done. *Transport Policy*, Vol. 9, No. 2, 2002, pp. 117–127. https://doi.org/10.1016/S0967-070X(02)00011-2.
25. Taylor, M. A. P., and G. M. D'Este. Transport Network Vulnerability: A Method for Diagnosis of Critical Locations in Transport Infrastructure Systems. In *Critical Infrastructure* (Murray, P. A. T., and P. T. H. Grubesic, eds.), Springer, Berlin, Heidelberg, Germany, 2007, pp. 9–30.
26. Jenelius, E., and L.-G. Mattsson. Road Network Vulnerability Analysis of Area-Covering Disruptions: A Grid-Based Approach with Case Study. *Transportation Research Part A: Policy and Practice*, Vol. 46, No. 5, 2012, pp. 746–760. https://doi.org/10.1016/j.tra.2012.02.003.
27. Knoop, V. L., M. Snelder, H. J. van Zuylen, and S. P. Hoogendoorn. Link-Level Vulnerability Indicators for Real-World Networks. *Transportation Research Part A: Policy and Practice*, Vol. 46, No. 5, 2012, pp. 843–854. https://doi.org/10.1016/j.tra.2012.02.004.
28. Briceño-Garmendia, C., H. Moroz, and J. Rozenberg. *Road Networks, Accessibility, and Resilience: The Cases of Colombia, Ecuador, and Peru*. The World Bank, Washington, D.C., 2015.
29. Rozenberg, J., H. E. Moroz, C. M. Briceno-Garmendia, X. Lu, and L. Bonzanigo. *Improving the Resilience of Peru's Road Network to Climate Events*. Publication WPS8013. The World Bank, Washginton, D.C., 2017, pp. 1–34.
30. Sampson, C. C., A. M. Smith, P. D. Bates, J. C. Neal, L. Alfieri, and J. E. Freer. A High-Resolution Global Flood Hazard Model. *Water Resources Research*, Vol. 51, No. 9, 2015, pp. 7358–7381. https://doi.org/10.1002/2015WR016954.
31. Wang, W., and F. Guo. RoadLab: Revamping Road Condition and Road Safety Monitoring by Crowdsourcing with Smartphone App. Presented at 95th Annual Meeting of the Transportation Research Board, Washington, D.C., 2016.
32. Espinet, X., W. Wang, and S. Mehndiratta. Low-Budget Techniques for Road Network Mapping and Road Condition Assessment That Are Accessible to Transport Agencies in Developing Countries. *Transportation Research Record: Journal of the Transportation Research Board*, 2017. 2634: 1–7.

The Special Task Force on Climate Change and Energy (A0020T) peer-reviewed this paper (18-00195).

Article

TRR
JOURNAL OF THE TRANSPORTATION RESEARCH BOARD

Transportation Research Record
2018, Vol. 2672(2) 54–67

DOI: 10.1177/0361198118797213
journals.sagepub.com/home/trr
SAGE

Urban Transportation Mode Choice and Carbon Emissions in Southeast Asia

Wei-Shiuen Ng[1]

Abstract

Cities are growing differently across the world, even within the same region, and presenting different transportation trends and challenges. Existing transportation services and travel behavior are some of the key variables shaping future transportation trends and carbon emissions projections. This study uses five developing cities in Southeast Asia to illustrate how different policy scenarios can help cities achieve more sustainable transportation development. Cities in Southeast Asia encompass distinctive characteristics, such as a wide range of transportation alternatives, often in the form of informal transit, and although they are not growing as rapidly as Chinese or Indian cities, their levels of transportation emissions have been increasing consistently. This study examines how different policies and measures will affect transportation mode choice and carbon emissions through the construction of mode choice models and the application of three policy scenarios. Carbon emissions can be reduced by as much as 93% in 2050 if cities implement a combination of land use planning changes, public transportation development, and economic policies for a modal shift to more energy efficient mode choices. Such policies and measures will therefore be able to contribute to city level climate goals or national climate targets.

According to the International Transport Forum (ITF) (*1*), 43% of the world's transportation demand in passenger kilometers will be in China, India, and the rest of Asia by 2050. This is a region that has been projected to grow significantly and rapidly in population size, economic development, urbanization rate, and motorization level. The ten countries that make up the Association of Southeast Asian Nations (ASEAN) have a combined population of 645 million and a total GDP of US$2.72 trillion (*2*), with a projected average annual economic growth of 5.1% over 2017 to 2021 (*3*). On average, the rate of motorization in the region, mostly in the form of two wheelers, increased by 12% from 2002 to 2010. During the same period, fuel consumption increased by 10%, with an over 10% growth for Indonesia and Vietnam and between 2% and 6% for the Philippines and Malaysia (*4*).

Although increases in motorization will bring positive benefits and contribute to economic growth, high levels of congestion, energy consumption, local air pollution, and carbon dioxide (CO_2) emissions will often follow. There is no doubt that the growth in vehicle ownership, transportation demand, and emissions in Chinese and Indian cities is unprecedented but Southeast Asian cities, despite growing at a smaller magnitude but just as rapidly, cannot be ignored. Southeast Asian cities' transportation demand and subsequent CO_2 emissions have been growing steadily and will contribute significantly to the region's projected transportation CO_2 emissions (*1*). This study focuses on urban transportation trends and CO_2 emissions in five cities in Southeast Asia, namely, Hanoi, Jakarta, Kuala Lumpur, Manila, and Phnom Penh. A set of three policy scenarios is applied to evaluate the impact of a range of policies and measures on mode choice and CO_2 emissions levels, after constructing a mode choice model for each city. A better understanding of effective policies and measures for different cities will help guide city level climate goals that could be transformed into national level commitments or targets.

Transportation Demand in Southeast Asia

Transportation demand in Southeast Asia has been increasing steadily over the past three decades and has not shown any signs of slowing down. The majority of the private vehicles owned are two wheelers, such as scooters, mopeds, and motorcycles. Recent statistics from the IEA (*5*) show that the number of two wheelers grew by 177% in Vietnam between 2000 and 2013. At the

[1]International Transport Forum, Paris, France

Corresponding Author:
Address correspondence to Wei-Shiuen Ng: wei-shiuen.ng@itf-oecd.org

same time, the growing middle class and rising household incomes have been driving urban motorization, partly because car ownership is a status symbol in Asia (*6*). The increase in light-duty vehicles (LDVs), which include cars, sports utility vehicles, and pick-up trucks, was 600% in Vietnam during the same period. In Malaysia, growth in LDVs was 148%, while the Philippines saw a relatively lower percentage at 44% between 2000 and 2013. Indonesia also experienced high vehicle growth rates for both two wheelers and LDVs, with increases of over 600% and 280% respectively. Most LDVs have also become single occupancy vehicles most of the time, leading to greater time delays and higher vehicular emission levels in urban areas. It is not surprising that due to their affordability and practicality, two wheelers are still the leading transportation mode choice in these countries. In Vietnam, 95% of all vehicles are two wheelers (*5*). These increases are a reflection of the lack of an adequate public transportation system and the general appeal of personal mobility over public transportation. In Kuala Lumpur, the poor state of public transportation services, as well as rising household incomes, have increased private vehicle ownership and decreased the use of public transportation systems simultaneously (*7*).

Transportation Externalities

Local air pollution from transportation activities and their consequent health impact, along with traffic congestion, are pressing challenges in many Southeast Asian countries. For example, traffic congestion costs Cambodia around US$6 million per month as a result of lower economic efficiency and the loss of working time and fuel (*8*). In Kuala Lumpur, the World Bank estimated the costs of traffic congestion at between 1.0% and 1.8% of the national GDP in 2014 (*9*). Congestion is further accentuated by the patterns of motorization and the predominance of motorcycles on the road in Southeast Asian cities. In Phnom Penh, disorderly use of the road by motorcycles, which comprise 70% of the city road traffic, fosters urban congestion and is responsible for traffic fatalities (*10*).

Although current transportation priorities in Southeast Asia are local air pollution and traffic congestion, this is a region that has increasingly become a larger contributor of global CO_2 emissions. The increasing motorization rate in Vietnam led to a 190% increase in its greenhouse gas (GHG) emissions from the transportation sector between 2000 and 2010, which is higher than China (160%) and India (100%) over the same period (*11*). Indonesia, Cambodia, and Malaysia also recorded significant increases of transportation-induced GHG emissions, at 63%, 46%, and 37% respectively between 2000 and 2010 (*11*). The Philippines was the only country selected in this study to see a slight decrease (3%) in its transportation GHG emissions. As GHG emissions become a growing concern in the region, there is now a greater sense of urgency for Southeast Asian cities to adopt more sustainable transportation development policies and measures, which are aligned with their national climate targets and commitments.

Current Studies

While current mode choice analyses for Southeast Asia often offer policy insights for cities (*7*, *12*, *13*), most studies have not tied mode choice with transportation planning or transportation demand projections for cities. Moreover, the policy measures recommended in the existing literature fail to evaluate the impact of combined policies on transportation demand and emissions. Most of the existing mode choice analyses are focused on the differences in motorization patterns between developed and developing Asian cities (*7*) and the impact of vehicle ownership trends on private vehicle use and preferences (*12*). Hyodo et al. (*14*) conducted an overview of urban travel behavior in 13 global cities, including Southeast Asian cities, but did not go beyond a comparison of descriptive statistics. Most of the current studies on transportation development in Southeast Asia are only focused on transit-oriented development as a solution for congestion, while comparative analyses have addressed different patterns of transportation development and urban sprawl, which is leading to longer travel distance (*15*, *16*). Densely populated cities in Southeast Asia provide an ideal urban landscape for the successful implementation of intensive public transportation policies. However, the abundance of informal transportation systems in Southeast Asian cities will hinder the development of public transportation unless they can be integrated into the transportation system (*17*).

Policies that support the adoption of advanced vehicle technology and alternative fuels, such as fuel efficiency standards, will reduce CO_2 emission by increasing energy efficiency. At the same time, it is just as critical to include non-technological measures to prevent the increase in private vehicle use and transportation demand, which could surpass the improvements made in energy efficiency. Therefore, a comprehensive range of policies and measures, including land use planning, public transportation development, economic instruments, and government regulations, need to be implemented for low-carbon transportation systems. This study created three policy scenarios that reflect a combination of such policies and evaluated their impact on transportation mode choice and CO_2 emissions in five cities in Southeast Asia, after the construction of a separate mode choice model for each city.

Table 1. Transportation Choice Set and Data

City	Sample size	Number of observations	Transportation alternatives in mode choice model
Hanoi	204,232	187,830	walk; bicycle; two wheeler (motorcycle); LDV (drive) (car); taxi; bus; three wheeler ("cyclo")
Jakarta	1,083,280	1,079,821	walk; bicycle; two wheeler (motorcycle); LDV (drive) (sedan, jeep, "kijang," pick-up truck); taxi; bus; train/rail; three wheeler ("bajaj," "becak")
Kuala Lumpur	218,460	216,527	walk; bicycle; two wheeler (motorcycle); LDV (drive) (car, small van, pick-up truck); taxi; bus; train
Manila	471,035	466,301	walk; bicycle; two wheeler (motorcycle); LDV (drive) (car, jeep); taxi; bus; train/rail; three wheeler (pedicab, tricycle)
Phnom Penh	40,368	39,013	walk; bicycle; two wheeler (motorcycle, "motodop," "motorumo"); LDV (drive) (passenger, pick-up truck, van); taxi; bus; three wheeler ("cyclo")

Methodology

This study applied discrete choice analysis to construct mode choice models that offer insights into travel behavior and policy, as well as to use the parameters derived from the choice models to conduct mode choice and CO_2 emission projections to 2050 under a mix of policy decisions. Household travel survey data, which included mode choice, travel distance, travel time, and socioeconomic variables, collected by the Japan International Cooperation Agency (JICA) for the cities of Hanoi, Jakarta, Kuala Lumpur, Manila, and Phnom Penh were used to estimate a disaggregate transportation mode choice model for each city. This enabled the evaluation of land use and transportation planning measures, pricing policies, and government regulations on mode choice and CO_2 emissions subsequently.

Disaggregate demand models can capture the variation in individual characteristics; they do not assume homogeneity among different consumers (*18*, *19*) and can include a large set of transportation attributes (*20*). Disaggregate models are sometimes also known as behavioral models because they depict individual travel choices (*21*) and can explain behavior directly at the level of a person, household, or firm. Since these models often analyze choices among discrete and not continuous alternatives, they are also known as discrete choice models. In this study, household trip level data were used to develop five multinomial logit models, which allowed the projections of 2050 travel demand and CO_2 emissions for five different cities.

Mode Choice Model

Transportation mode choice is analyzed in this study using a multinomial logit (MNL) model, which is a commonly used form of discrete choice analysis (*22*, *23*). The outcomes of MNL models are logit choice probabilities of each alternative as a function of the systematic portion of the utility of all the alternatives available in a choice set. In this study, a separate MNL model was constructed for the mode choice of each city.

Since the data collected differ across the five cities, each choice model is slightly different. The final choice set for each city contains at least seven alternatives (Hanoi, Kuala Lumpur, and Phnom Penh), while the other two cities have eight alternatives. These alternatives were derived from a much larger group of alternatives from the original surveys that ranged from 13 (Phnom Penh) to 21 (Jakarta) alternatives, some of which specified different types of LDVs, two wheelers, or three wheelers, as well as informal transit services, most of which have been combined and recoded into one of the final seven or eight alternatives as presented in Table 1.

Table 1 also shows the number of observations for each choice model, which is different from the original sample size for each city because incomplete observations were removed from the final models. In addition, only choices that were available to each household (as indicated in the household travel surveys), either in the form of ownership or use for LDVs, two wheelers, and bicycles, were included in the final mode choice models.

Utility Functions

The utility of each alternative in the choice set is expressed as follows, where U_{in} is the utility of the *i*th alternative for the *n*th individual:

$$U_{in} = \alpha_i + \beta_i X_{in} + \varepsilon_{in} \tag{1}$$

where α_i is the alternative specific constant (ASC), β_i is the vector of unknown parameters, X_{in} represents the vectors of known variables (e.g. travel time, travel cost, and socioeconomic variables), while ε_{in} is the unobserved error term.

An example utility function for LDV used in the model specification for Hanoi is shown below in Equation 2:

$$\begin{aligned} U_{LDV} = {} & \alpha_{LDV} + \beta_{TRAVEL\,TIME} LDV\ TRAVEL\ TIME \\ & + \beta_{TRAVEL\,COST} LDV\ TRAVEL\ COST \\ & + \beta_{HOUSEHOLD\,SIZE} HOUSEHOLD\ SIZE \\ & + \beta_{HOUSEHOLD\,INCOME} HOUSEHOLD\ INCOME \\ & + \beta_{AGE} AGE + \beta_{GENDER} GENDER \end{aligned} \tag{2}$$

The "walk" alternative was normalized to zero in all five mode choice models, which implies that this alternative was used as a base with a zero constant and other constants would therefore be interpreted as relative to choosing to "walk" as a primary transportation mode.

All of the five mode choice models included travel time, travel cost, and household income, which are key variables that have been shown to affect mode choice. Household size, age, and gender variables were also included for most of the cities. Data on age and gender were unavailable for Phnom Penh, while the survey for Jakarta did not include household size data.

Travel Time, Distance, and Cost Estimations

Since most trips usually consist of multiple modes and unless the primary mode choice for each observation was already indicated on the survey (e.g., Phnom Penh), a primary mode choice was selected to represent the mode used to travel the longest distance. The mode with the maximum distance traveled was estimated when the travel time for each mode within the same trip was provided in the survey data (Hanoi and Kuala Lumpur). Travel distance was then estimated using an average speed for each mode. If travel time (or distance) per each mode segment was unavailable or could not be estimated, a mode hierarchy was applied (for Jakarta and Manila). This method was applied because of the lack of detailed origin and destination locations which prevented the geocoding of latitude and longitude coordinates. In the mode hierarchy, the selection of the primary mode choice is based on the order: rail public transport, bus, taxi, drive (LDV), motorcycle (or two wheeler), bicycle, three wheeler, and walk.

Travel time was estimated for the primary mode choice and the alternatives using the departure and arrival time data when travel time data were not specified in the survey. In fact, only Phnom Penh's survey provided travel time data for the primary mode choice. Travel cost was also estimated for each primary mode choice and every alternative, except for walk and bicycle choices. Travel cost was estimated for two wheelers and LDVs by multiplying trip distance with fuel cost and fuel economy, while LDV also included the addition of parking cost, that is ((distance * fuel cost * fuel economy) + parking cost). Travel cost for taxis was estimated by multiplying trip distance and per km cost and the addition of a fixed cost which differs by city. Categorical variables were created for bus and three wheeler costs, which may be dependent upon travel distance, based on transit fares and pricing structure in each city.

Transportation Demand and CO_2 Emission Forecasts in Three Policy Scenarios

Discrete choice models have the ability to forecast future behavior under different circumstances as captured by the variables included in the models. The outcomes (parameter estimates) of the MNL models in this study were used as inputs to estimate changes in travel behavior and CO_2 emissions under three scenarios in 2050.

The outcomes of MNL models are logit choice probabilities of each alternative as a function of the systematic portion of the utility of all the alternatives available in a choice set. The model assumes that each individual chooses an alternative based on the theory of maximum utility, where the maximum likelihood estimation is applied (*22*). Since the utility function is assumed to be linear in parameters (β), implying $V_{nj} = \beta X_{nj}$, where β is a row vector of unknown parameters and X_{nj} is a column vector of observed variables associated with alternative j, the logit probabilities can be expressed as Equation 3:

$$P_{ni} = \frac{e^{\beta X_{ni}}}{\sum_j e^{\beta X_{nj}}} \tag{3}$$

The logit probability of each alternative was estimated using the parameter (β) derived from the choice models and the constructed utility functions such as Equation 2.2. The next step was then to develop aggregate forecasts using the logit probabilities from the disaggregate models. The classification aggregation method (*24*, *19*) was applied in this study by grouping the sample population according to household income (up to 14 groups). Disaggregate probabilities by income were estimated by dividing the sum of the product of choice probability and expansion factor or weight (provided in the household travel survey datasets) by the sum of expansion factor. Changes in household income over time, up to 2050, were modeled upon the GDP growth rate of each city for the same period of time. A new set of disaggregate mode share for each city was then derived using the estimated weighted probability values.

Transportation Behavior and Demand Estimation. The annual passenger kilometer (pkm) was estimated based on the disaggregate choice probability for each alternative, multiplied by the population size (*25*), average trip distance, daily trip rate, and 365 days. Average trip distance was estimated based on the elasticity of trip distance with respect to population, which was assumed to be 0.4555 (*1*). The average trip distance for the base year was derived from the survey data for each mode in each city. The average trip rate (number of trips per day per person) was assumed to be a function of GDP per capita, as shown in Equation 4, with the parameters estimated from a regression analysis using regional data for Southeast Asia:

$$Trip\,Rate = 2.09 + (0.0193 * (GDP\,per\,Capita)) \quad (4)$$

The vehicle kilometer (vkm) was estimated by multiplying the pkm with load factor for two wheeler, LDV, bus, taxi, and three wheeler alternatives. The load factors used in this study were based on the IEA's assumptions (*5*).

Transportation Energy and CO_2 Emission Estimation. Energy use was estimated for each mode based on the IEA's assumptions (*5*), which include a detailed characterization of vehicle technology and fuel mix, including internal combustion engines that run on gasoline, diesel, compressed natural gas, or liquid petroleum gas, hybrid electric, plug-in hybrid electric, electric, and fuel cell powertrains. Related fuel options include: gasoline and diesel, biofuel, and synthetic fuel alternatives to liquid fuels, gaseous fuels including natural gas and hydrogen, and electricity. Total energy use (*E*) for vehicle type *i* can therefore, be estimated using the following equation:

$$E_i = \sum (N_i * I_{i*} D_i) \quad (5)$$

where E_i is the total energy use, N_i is the total number of vehicle type *i*, I_i is the energy intensity for vehicle type *i*, which also reflects fuel type, and D_i is the average distance traveled by vehicle type *i*. Total CO_2 emissions were then estimated for each fuel type using coefficients of carbon per unit of energy consumed in vehicle type *i*.

Policy Scenarios. There are three policy scenarios designed for each city, which reflect a range of policies and measures, such as land use planning, public transport development, economic instruments, and government regulations that have been shown to influence transport demand and emissions. In the Baseline scenario, it is assumed that there will be no implementation of significant policies and measures. The underlying factors driving transportation demand, such as population, household income, trip distance, and trip rate will continue to grow but there will be no changes in mode share or travel time and cost. Load factors for LDV and bus were assumed to decrease by 10% from 2015 to 2030 and by 27% in 2050. In the second scenario, Robust Governance (ROG), local governments will play a more significant role by introducing economic instruments, such as parking pricing, road tolls, and higher subsidies for public transportation. The third scenario, Integrated Land Use and Transport Planning (LUT), assumes the same set of economic instruments shown in Robust Governance but coupled with changes in land use and transportation planning that will decrease public transportation travel time, as well as an average trip distance that is equivalent to the 2015 level. In all three scenarios, the fuel economy standards, which reflect fuel mix and market penetration of different vehicle technologies, will follow the IEA's assumptions (*5*). Apart from the Baseline, the other two scenarios follow a stringent set of fuel economy standards that will lead to a global CO_2 emission reduction of almost 60% between 2013 and 2050.

The scenario outcomes are not predictions, but different possible futures based on the assumptions applied in each scenario. The impacts of the policies and measures in each scenario are evaluated by changing the relevant components (e.g., travel time and cost) in the utility functions based on the assumptions indicated in the scenarios (Table 2). Probabilities (Equation 3) for all the alternatives were then calculated again under the three different scenarios.

Mode Choice Model Estimation Results

The model estimation results of transportation mode choice in five Southeast Asian cities are presented in Table 3. A separate mode choice was estimated for each city. The most preferred mode choice (as indicated by the constant parameter estimates) is different for each city. Other than Phnom Penh, where two wheeler is the most preferred mode choice, walking seems to be most preferable in the other four cities. This is quite unusual but could be due to the high level of traffic congestion that affects the other modes and the relatively short travel distances. Bicycle is the second most preferred choice in Hanoi, followed by two wheeler and LDV. Public transportation in Hanoi is not as preferable as personal transportation, which provides a greater sense of freedom and flexibility. Bus ridership in Hanoi has always been low, constituting just 3% of the total mode share. In fact, the mode share for two wheeler was 12 times higher than bus, including both mini and standard buses. Despite increases in the share of bus mode over the past decade, recent data have shown declining bus ridership once again (*26*). Similarly, bus is not a preferred mode choice in Phnom Penh, which is another city with a high share

Table 2. Key Assumptions for Policy Scenarios

Scenario	Baseline		ROG		LUT	
	2030	2050	2030	2050	2030	2050
Public transport development						
Bus travel time[a] decrease (%)	-	-	-	-	30	60
Rail travel time[a] decrease (%)	-	-	-	-	30	60
Economic instruments						
Fuel tax increase[b] (%)	-	-	33	63	33	63
Parking pricing increase (%)	-	-	80	160	80	160
Road tolls (US$)[c]	-	-	1.02	1.36	1.02	1.36
Bus subsidy increase (%)	-	-	30	50	30	50
Rail subsidy increase (%)	-	-	30	50	30	50
Government regulations						
Fuel economy standards[d]	IEA 4DS	IEA 4DS	IEA 2DS	IEA 2DS	IEA 2DS	IEA 2DS

[a]Based on household travel survey data from each city. In the Baseline and ROG scenarios, the travel time for all modes used in the mode choice and emission models were all from the household travel survey data. In the LUT scenario, travel time for bus and train would decrease by 30% in 2030 and by 60% in 2050.
[b]Tax increase rates are based on Korea and Japan's fuel tax percentages.
[c]Road tolls, which are uncommon in Southeast Asian cities, are assumed to be implemented in all cities in the ROG and LUT scenarios.
[d]Fuel economy standards reflect the type of fuel mix and are based on IEA's assumptions in the 4°C Scenario (4DS) and the 2°C Scenario (2DS). The 4°C Scenario (4DS) takes into account countries' current intentions to limit emissions and improve energy efficiency that will limit the long-term temperature increase to 4°C. The 2DS reduces CO_2 emissions by almost 60% by 2050 (compared with 2013) and a projection of declining carbon emissions post 2050 until carbon neutrality is reached (5).

of two wheeler mode. However, public transportation (bus) is the second most preferred mode choice in Jakarta and Manila. Bus as a primary mode choice is also prefeered over most other modes in Kuala Lumpur. Such differences across these three cities are due to the disparities in the level of bus frequency and network size. The bus alternative in all choice sets also includes informal bus services, where applicable. Such bus services can be more convenient than other modes, especially for lower household income groups and for cities without a well-connected transportation system. None of the five cities, apart from Phnom Penh, showed driving (i.e., LDV use) as a relatively highly preferred mode choice. LDV is also not as preferable as two wheeler use in Hanoi, Jakarta, Kuala Lumpur, and Phnom Penh. However, in Manila, two wheeler use is the least preferred mode across all eight alternatives (Table 3).

The parameter estimates for travel time for all the five cities are negative and highly significant, as the longer the travel time, the less attractive the alternatives will be. Since it was assumed that the impact of time on preferences would be equal across all alternatives, the parameter for travel time was constrained to be the same for all alternatives in all five cities. On the other hand, the parameter for travel cost was only constrained across all alternatives for Hanoi, Jakarta, and Manila, which is a common practice for mode choice models which assume that travel cost affects all alternatives the same way. Different parameters were used for the travel costs of two wheeler, LDV, taxi, bus, train, and three wheeler in the choice models for Kuala Lumpur and Phnom Penh because in these two cities it was found that travel cost affects the utility of each mode differently. Travel costs of taxi and bus alternatives were in fact insignificant for respondents in Phnom Penh. The travel cost parameter estimates for all the other four cities are negative and significant (Table 3).

The socioeconomic characteristics, such as household size, household income, age, and gender of the respondents were also included in the five mode choice models as shown in Table 3.

The impact of household size on mode choice does not follow a clear trend across the cities. It is a significant variable for most modes in most cities, but it is especially insignificant for taxi in Hanoi and Kuala Lumpur, train in Kuala Lumpur (but not in Manila, where the larger the household size, the less attractive the train mode becomes), and two wheeler, LDV, bus, and three wheeler use in Phnom Penh.

Household income is highly significant for all modes in Manila, all modes but train in Kuala Lumpur and Jakarta, and all modes but bicycle in Phnom Penh. Household income is also insignificant for bicycle use in Hanoi. The use of two wheeler becomes more attractive when household income rises for all cities except in Jakarta, where rising income will decrease its appeal instead. This is also true for LDV, where rising income will increase its utility in all cities but Jakarta. In Hanoi and Kuala Lumpur, the utility for bus mode will decrease when household income increases, while in Manila and Phnom Penh taxi becomes more attractive when household income increases.

Table 3. MNL Models Estimation Results of Transportation Mode Choices in Five Cities

Explanatory variables	Hanoi			Jakarta			Kuala Lumpur			Manila			Phnom Penh		
	Parameter estimates	T-test	*P*-value	Parameter estimates	T-test	*P*-value	Parameter estimates	T-test	*P*-value	Parameter estimates	T-test	*P*-value	Parameter estimates	T-test	*P*-value
Walk constant	-	-	-	-	-	-	-	-	-	-	-	-	-	-	-
Bicycle constant	−1.77	−49.39	0.00	−4.28	−213.56	0.00	−4.98	−71.18	0.00	−6.42	−76.44	0.00	−8.79	−9.03	0.00
Two wheeler (2W) constant	−5.77	−101.21	0.00	−5.53	−330.42	0.00	−6.85	−106.97	0.00	−14.50	−79.01	0.00	0.60	15.69	0.00
LDV constant	−6.70	−78.31	0.00	−5.22	−365.73	0.00	−5.95	−95.31	0.00	−10.10	−154.36	0.00	−1.17	−8.24	0.00
Taxi constant	−19.90	−41.92	0.00	−4.36	−254.85	0.00	−0.17	−0.04	0.97	−7.77	−122.01	0.00	−6.55	−15.14	0.00
Bus constant	−9.15	−17.08	0.00	−1.33	−216.77	0.00	−3.05	−11.89	0.00	−0.93	−65.58	0.00	−6.23	−0.75	0.45
Train constant	-	-	-	−6.46	−237.65	0.00	−8.66	−25.36	0.00	−6.95	−124.44	0.00	-	-	-
Three wheeler (3W) constant	−13.00	−8.45	0.00	−4.46	−182.47	0.00	-	-	-	−1.05	−52.59	0.00	−3.10	−22.88	0.00
Household size - bicycle	−0.03	−6.45	0.00	-	-	-	0.05	6.67	0.00	0.01	1.24	0.22	−0.22	−13.94	0.00
Household size - 2W	−0.05	−8.16	0.40	-	-	-	-0.03	−4.93	0.12	0.02	0.71	0.48	0.01	1.62	0.11
Household size - LDV	−0.14	−11.98	0.00	-	-	-	-0.04	−7.74	0.00	−0.01	−1.15	0.25	0.01	1.20	0.23
Household size - taxi	0.39	10.06	0.00	-	-	-	-0.83	−3.34	0.00	−0.15	−19.18	0.00	0.11	3.31	0.00
Household size - bus	0.18	3.45	0.00	-	-	-	0.02	3.98	0.00	−0.06	−27.66	0.00	0.03	0.57	0.57
Household size - train	-	-	-	-	-	-	0.07	1.54	0.00	−0.06	−7.27	0.00	-	-	-
Household size - 3W	−0.16	−0.83	0.00	-	-	-	-	-	-	−0.02	−6.01	0.00	0.02	1.38	0.17
Household income - bicycle	0.00	−1.20	0.00	−0.01	−3.76	0.00	0.00	−6.00	0.00	−0.12	−9.20	0.00	0.36	0.96	0.34
Household income - 2W	0.04	20.11	0.07	−0.04	−15.63	0.00	0.00	5.48	0.63	0.21	9.66	0.00	0.19	17.04	0.00
Household income - LDV	0.03	10.75	0.03	−0.04	−16.73	0.00	0.00	−3.61	0.00	0.61	78.98	0.00	0.39	21.81	0.00
Household income - taxi	−0.12	−2.23	0.00	−0.01	−2.45	0.01	−0.77	−4.09	0.00	0.46	66.55	0.00	0.69	16.58	0.00
Household income - bus	−0.45	−5.76	0.00	0.00	2.22	0.03	0.00	−8.95	0.00	0.15	56.61	0.00	0.19	2.72	0.01
Household income - train	-	-	-	0.00	0.42	0.68	0.00	−0.48	0.00	0.31	40.58	0.00	-	-	-
Household income - 3W	−0.15	−1.79	0.00	−0.01	−2.25	0.02	-	-	-	−0.02	−5.86	0.00	0.33	12.39	0.00
Age - bicycle	−0.02	−52.71	0.00	0.01	22.55	0.00	−0.01	−7.30	0.00	0.03	34.14	0.00	-	-	-
Age - 2W	0.00	−9.32	0.00	0.03	107.13	0.00	0.01	11.73	0.00	0.01	4.77	0.00	-	-	-
Age - LDV	−0.03	−22.25	0.00	0.03	125.56	0.00	0.03	20.04	0.00	0.04	45.94	0.00	-	-	-

(continued)

Table 3. (continued)

Explanatory variables	Hanoi Parameter estimates	Hanoi T-test	Hanoi *P*-value	Jakarta Parameter estimates	Jakarta T-test	Jakarta *P*-value	Kuala Lumpur Parameter estimates	Kuala Lumpur T-test	Kuala Lumpur *P*-value	Manila Parameter estimates	Manila T-test	Manila *P*-value	Phnom Penh Parameter estimates	Phnom Penh T-test	Phnom Penh *P*-value
Age - taxi	−0.12	−51.76	0.00	0.02	55.51	0.00	−2.41	−4.71	0.01	0.03	46.75	0.00	-	-	-
Age - bus	0.03	4.36	0.00	0.00	13.70	0.00	−0.02	−15.87	0.00	0.01	46.07	0.00	-	-	-
Age - train	-	-	-	0.02	36.56	0.00	0.02	2.78	0.00	0.00	5.19	0.00	-	-	-
Age - 3W	−0.41	−53.11	0.23	0.02	31.60	0.00	-	-	-	0.00	1.97	0.05	-	-	-
Gender - bicycle	−0.15	−11.14	0.00	0.54	39.17	0.00	0.50	16.21	0.14	1.53	33.54	0.00	-	-	-
Gender - 2W	0.76	39.18	0.00	0.75	72.16	0.00	1.50	61.88	0.88	2.00	21.78	0.00	-	-	-
Gender - LDV	0.43	13.32	0.00	0.51	52.07	0.00	0.15	6.72	0.00	0.83	32.55	0.00	-	-	-
Gender - taxi	1.63	4.50	0.00	−0.09	−8.05	0.00	−0.09	−0.15	0.00	0.20	8.68	0.00	-	-	-
Gender - bus	1.58	5.08	0.00	0.04	10.22	0.00	−0.03	−1.49	0.00	−0.03	−3.84	0.00	-	-	-
Gender - train	-	-	-	0.47	25.30	0.00	−1.06	−4.78	0.00	0.12	4.38	0.00	-	-	-
Gender - 3W	0.43	1.97	0.05	−0.53	−27.10	0.00	-	-	-	−0.13	−13.04	0.00	-	-	-
Travel cost (US$)	−4.03	−31.31	0.00	−0.08	−56.24	0.00	-	-	-	−0.26	−37.74	0.00	-	-	-
Travel cost - 2W (US$)	-	-	-	-	-	-	−20.10	−50.98	0.00	-	-	-	−29.40	−16.75	0.00
Travel cost - LDV (US$)	-	-	-	-	-	-	−9.19	−44.36	0.00	-	-	-	−53.30	−12.56	0.00
Travel cost - taxi (US$)	-	-	-	-	-	-	−0.61	−6.40	0.00	-	-	-	−0.04	−1.25	0.21
Travel cost - bus (US$)	-	-	-	-	-	-	−1.67	−4.12	0.00	-	-	-	−2.29	−0.10	0.92
Travel cost - train (US$)	-	-	-	-	-	-	−5.81	−22.99	0.00	-	-	-	-	-	-
Travel cost - 3W (US$)	-	-	-	-	-	-	-	-	-	-	-	-	−0.22	−5.77	0.00
Travel time (min)	−0.23	−122.55	0.00	−0.03	−493.22	0.00	−0.26	−74.54	0.00	−0.06	−196.39	0.00	−0.02	−13.33	0.00
Summary statistics															
Number of observations	187,830			1,079,821			216,527			466,301			39,013		
Rho square	0.44			0.20			0.28			0.24			0.29		

Note: Missing data denote parameters that are unavailable due to model constraints or variables not included in the original surveys.

Table 4. Value of Time Estimates

Value of time	Hanoi	Jakarta	Kuala Lumpur	Manila	Phnom Penh
Constrained (US$/hr)	3.45	22.12		14.20	
2W (US$/hr)			0.79		0.04
LDV (US$/hr)			1.72		0.02
Taxi (US$/hr)			26.14		30.00
Bus (US$/hr)			9.49		0.52
Train (US$/hr)			2.73		-
3W (US$/hr)			-		5.45

Note: The values of time for three wheeler in Kuala Lumpur and for train in Phnom Penh are not estimated because these modes were unavailable in the respective city and, hence, excluded from its respective mode choice model.

Age is highly significant for all modes in all cities, except for taxi in Hanoi and Kuala Lumpur. Although age is a significant variable, its impact on different modes across the cities varies. For example, older respondents tend to find bicycle more attractive as a primary mode choice in Jakarta and Manila, but the opposite is the case in Hanoi and Kuala Lumpur. The utility of two wheeler remains high among older respondents in all cities, where age does not seem to be a concern when using two wheelers. Older respondents also tend to find LDV more attractive in all cities but Hanoi.

The parameter estimates for gender, where male = 1 in the dummy variable, are positive for all cities in the two wheeler and LDV alternatives. This implies that men choose two wheeler and LDV modes as their primary mode choice more than women. Women choose to bicycle more than men in Hanoi, while the opposite is true in Jakarta, Kuala Lumpur, and Manila. Women also choose to take taxis more than men in Jakarta and Kuala Lumpur, and they ride buses more than men in Kuala Lumpur and Manila. Kuala Lumpur is the only city where women found train to be a more attractive mode choice than men. In Jakarta and Manila, female travelers were also found to choose three wheelers more than male travelers (Table 3). Public transportation modes are, therefore, not always more attractive for women.

The parameters estimated from the mode choice models were used to calculate the value of travel. The value of travel time depends on the utility that an individual attaches to time spent in a particular mode and the opportunity cost of travel time (*27*). A high value of time can be the result of a high opportunity cost of time or a high level of disutility of time spent on a transportation mode (*20*). The value of time is also the marginal rate of substitution of time for cost ($MRS_{\text{Time-Cost}} = \frac{\beta_{Travel\ Time}}{\beta_{Travel\ Cost}}$), which expresses the willingness to pay for a specific transportation mode and the trade-off between time and cost while still maintaining the same level of utility. This value is different for each individual as it depends on observed travel time and wage rate and could change as trip purpose changes. The value of time can also change according to the characteristics of the traveler, time of travel, and trip duration (*28*). Table 4 shows the value of time estimates in the five different cities. Travelers in Jakarta have the highest value of time of US$22 per hour, compared with travelers in Hanoi and Manila, while taxi users have exceptionally high values of time in Kuala Lumpur and Phnom Penh (Table 4). The values of time of other modes apart from taxi in Phnom Penh are significantly lower than the other four cities. This is due to the relatively low average wage in Phnom Penh, which was US$ 0.44 per hour in 2011 (*29*) and is most likely much lower when the household travel survey was collected. Since the value of time is usually associated with the average wage rate (*30*, *28*), the differences in the value of time among the cities thus reflect different levels of wages.

Policy Scenario Results

The parameters derived from the mode choice models were used to estimate the probabilities of each mode (Equation 2.3), which were then used to forecast changes in future mode choice under three different policy scenarios, as described in the Policy Scenarios subsection. Since the household travel survey data were collected from different years, ranging from 1996 to 2005 depending on the city, the base year of 2015 was selected to standardize the estimations and each set of probability estimations was then calibrated accordingly, based on the expansion rate by household income growth (up to 2050). This section presents transportation demand in passenger kilometer (pkm) (Figure 1), vehicle kilometer (vkm) (Figure 2) and CO_2 emissions (Figure 3) in three scenarios in 2050.

As shown in Figure 1, pkm is the lowest in the LUT scenario for all five cities, though the magnitude of change for each city varies. The pkm levels in the Baseline and ROG scenarios are very similar for Hanoi, Jakarta, and Manila, while the greatest difference in

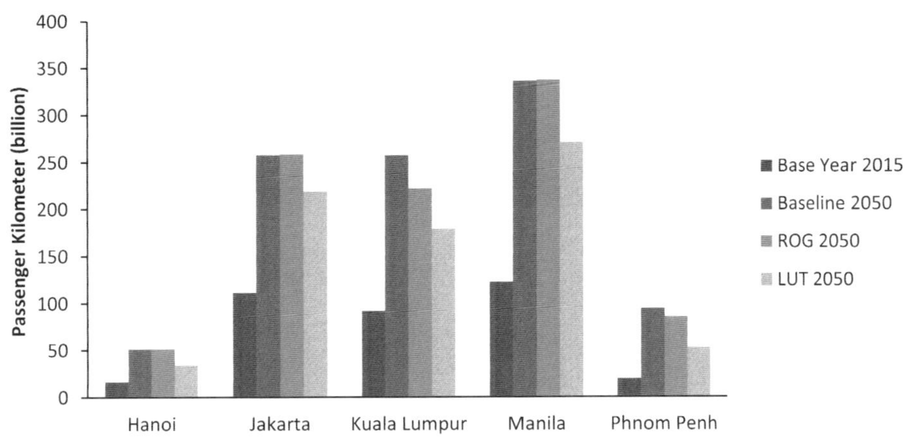

Figure 1. Total transportation demand in passenger kilometer (pkm) under three policy scenarios for five cities in 2050.

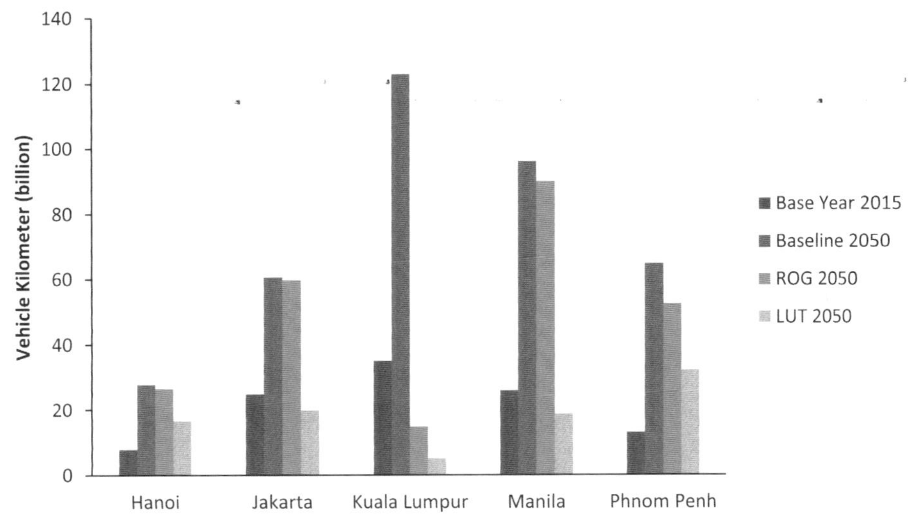

Figure 2. Total transportation demand in vehicle kilometer (vkm) under three policy scenarios for five cities in 2050.

pkm between the Baseline and ROG scenarios can be found in Kuala Lumpur (14%). This implies that the pricing policies implemented in the ROG scenario are not as effective in regulating transportation demand in the three former cities as they are in Kuala Lumpur and Phnom Penh. Even in the LUT scenarios, which have more stringent policy assumptions, pkm levels in 2050 are projected to be higher than in the 2015 base year in all cities.

In general, total pkm will increase from the base year under all three scenarios, ranging between 95% (Kuala Lumpur in LUT scenario) and almost 400% (Phnom Penh in Baseline scenario) in 2050. These trends are consistent with economic growth projections for the five cities. As a result of the integrated planning assumptions in LUT, pkm will increase at a lower rate from the base year compared with the Baseline and ROG scenarios because of changes in travel distance in all cities.

When converted to vkm, transportation demand is significantly lower in LUT than in the Baseline and ROG scenarios across all cities, with Jakarta, Kuala Lumpur, and Manila having lower vkm levels in the LUT scenario in 2050 than in the 2015 base year (Figure 2). This decline in vkm is due to the increase in public transport ridership

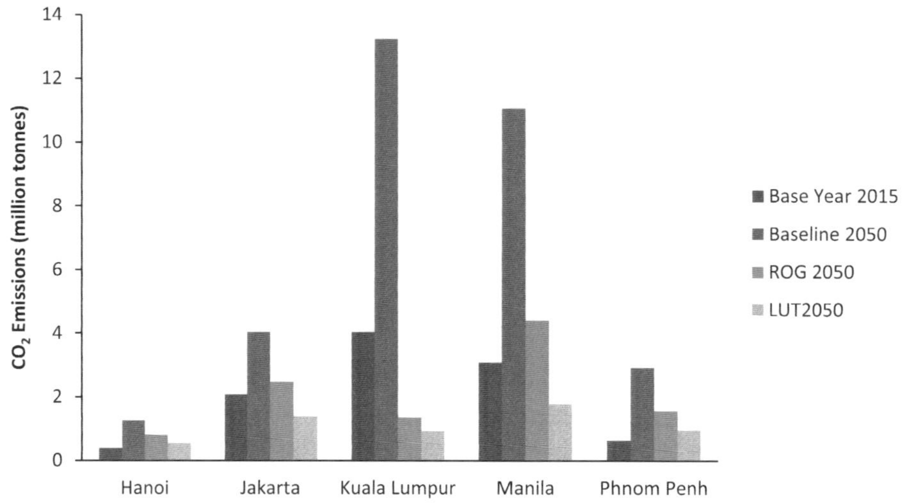

Figure 3. Total CO_2 emissions under three policy scenarios in 2050.

triggered by improvements in the quality of train and bus services and the reduction in their travel time. The difference in vkm between the Baseline and ROG scenarios can be as high as 88%, as seen in Kuala Lumpur, or as low as 1% in Jakarta. The wide range of impact reflects the varying influence different pricing policies have on mode choice and different preferences in mode choice in different cities. For example, the pricing policies assumed in the ROG scenario are more effective in regulating vkm in Kuala Lumpur than in Hanoi or Jakarta. The significant decrease in vkm in both the ROG and LUT scenarios in Kuala Lumpur is due to a substantial modal shift from LDVs (or private vehicles) to public transport. This shows that there can be another vkm growth alternative for Kuala Lumpur, one that does not rely heavily on the use of private vehicles as indicated in the Baseline scenario.

Integrated land use and transport planning assumptions seem to be most effective in Manila as vkm decreased by 80% from the ROG to LUT scenario, while land use measures have a smaller impact in Hanoi and Phnom Penh (both with a 38% decrease in vkm in LUT 2050), whose dominating mode choice is two wheeler even in 2050. The low share of public transportation in these two cities is the key reason why vkm in LUT 2050 remains higher than vkm in the 2015 base year, despite the implementation of robust policies and measures. Without better transportation alternatives, two wheelers will continue to dominate mode share in Hanoi and Phnom Penh.

Total CO_2 emission levels vary by city but, following vkm projection trends, they are lower in the LUT 2050 scenario than in the 2015 base year for Jakarta, Kuala Lumpur, and Manila (Figure 3). The policy scenarios have triggered the largest decrease in CO_2 emissions from the Baseline to LUT scenario in 2050 for Kuala Lumpur (93%), followed by Manila (84%). The difference in CO_2 emission reduction between the LUT and ROG scenarios ranged between 30% (Kuala Lumpur) and 60% (Manila), which showed that the impact of the same combination of policy measures on CO_2 emissions will vary depending on a city's transportation mode choice, preferences, existing and potential transportation alternatives, and fuel efficiency of vehicle fleet. With appropriate policies and measures, CO_2 emissions can be significantly reduced from the Baseline scenario by 2050, as even in Hanoi, where the smallest emission reduction was observed, the emission level in LUT will still be 58% lower than in the Baseline scenario. A low-carbon transportation future in Southeast Asian cities requires the implementation of specific pricing policies and land use measures that will target their distinctive mix of vehicle type, transportation services, and transit ridership.

Discussion and Conclusions

Cities are growing differently across the world, even within the same region, where there are different transportation trends and challenges. Existing transportation services and travel behavior are some of the key variables shaping future transportation trends and carbon emissions projections. Cities in Southeast Asia encompass distinctive characteristics, such as the wide range of transportation alternatives, often in the form of informal transit and two wheelers, including scooters, mopeds, and motorcycles. The most pressing transportation challenge experienced by Southeast Asian cities is not CO_2 emission but congestion or the lack of efficient public transportation services, which could lead to extended travel time delays and indirectly result in additional local air pollution and global CO_2 emissions. The abundance of informal public transportation services and high two wheeler use could hinder development of public transportation as they serve as alternative transportation services that are often perceived to be more flexible and convenient. For the cases of Hanoi and Phnom Penh, where two wheelers dominated the transportation mode share in 2015 (55% and 74% respectively) and will continue to do so in the LUT 2050 scenario (56% and 68% respectively), it would be difficult to shift two wheeler users to public transportation, as the freedom of mobility associated with two wheelers is the preferred choice. The utility of two wheelers in Hanoi and Phnom Penh is clearly higher than public transportation modes (Table 3). The quality of public transportation services, including shorter travel time, an extended network, and higher service frequencies, would have to be significantly improved in order to gain higher levels of ridership and reduce emissions in the long term. A cleaner two wheeler fleet composed of electric vehicles could also directly reduce local air pollution and CO_2 emissions.

Due to the high share of two wheelers in Hanoi, Jakarta, Kuala Lumpur, and Phnom Penh, any economic instruments implemented should always apply to both LDVs and two wheelers. Such instruments, including road tolls and parking policies, will then serve as transportation demand management tools for both types of vehicle. In addition, the high share of two wheelers reflects low public transportation and LDV mode shares. If cities were to regulate the use of two wheelers, the quality of public transportation would have to be improved at the same time to offer good alternatives.

Public transportation ridership, especially bus services, is projected to increase in Manila, Jakarta, and Kuala Lumpur, where public transportation will become the dominating mode choice in the LUT scenario in 2050. Public transportation ridership, including informal transit services, was already high in Manila and Jakarta in 2015. However, without robust pricing policies and land use measures to reduce the growth of private vehicle use, LDV mode share in Manila could increase to 28% in Baseline 2050 from 16% in 2015. The mode choice estimations (Table 3) show that LDV use in Manila is less attractive than public transportation, both bus and train. Hence, with further improvements made to its public transportation services, coupled with a range of policies and measures implemented to regulate private vehicle use, LDV mode share can be reduced to 2% in the LUT scenario in 2050, which explains the significant decrease in CO_2 emissions (Figure 3).

Kuala Lumpur is the only city in this study with a dominating LDV mode share in 2015 (46%) and Baseline 2050 (51%). It is also the city with the largest reduction in CO_2 emissions due to the changes in mode share triggered by different policies and measures, as illustrated in the policy scenarios. In both ROG and LUT 2050, bus and train mode shares will dominate and increase substantially to over 80% from 20% in Baseline 2050. This is a city with a high potential for massive public transportation development and improvement.

A key barrier to public transportation development in Southeast Asian cities is the wide range of informal public transportation services. Such services, which are also common in Latin American and African cities, provide more flexible route options and often at a lower cost than structured public transportation. High public transportation mode share can only be achieved when the provision of informal public transportation services is first addressed.

As shown in Table 4, the value of time estimates are relatively higher for bus and train when compared with LDV and two wheeler in Kuala Lumpur and Phnom Penh. In Jakarta, which has the highest constrained value of time across all modes, transportation users will be willing to pay more to spend less time traveling. Hence, the improvement of public transportation quality by reducing travel time will be more effective in increasing its attractiveness than reducing cost through transit subsidies.

A low-carbon transportation sector is possible to achieve even for rapidly growing Asian cities, as the increase in pkm does not necessarily imply a subsequent increase in vkm and CO_2 emissions, as seen in Jakarta, Kuala Lumpur, and Manila. Transportation demand in pkm can continue to grow but in a more sustainable manner that will keep vkm and CO_2 emissions low. The policy scenarios in this study illustrate that achieving low-carbon mobility requires targeted policies, which will differ by city according to the varying transportation preferences, constraints, and needs found in different cities. Pricing policies, such as fuel tax, parking, and road pricing, and bus and rail transit subsidies, are more effective in regulating pkm and vkm in Kuala Lumpur than in other cities. This is likely due to the high LDV mode share in Kuala Lumpur, which responds to the increased cost of driving and its greater bus utility (than LDV) which made it a more attractive mode when bus travel cost decreased while LDV travel cost increased. Having a set of robust pricing policies would therefore be an optimal way for Kuala Lumpur to reduce its transportation CO_2 emissions and improve its transportation services through additional pricing revenue. On the other hand, land use measures that will reduce average travel distance and transit travel time tend to have a greater impact on reducing transportation demand in Manila than in other cities. The diverse mix of transportation modes, including informal transit and three wheelers, will continue to provide services though not as much as in the base year. Land use measures, such as mixed use planning and transit-oriented development in neighborhoods that already have high density levels, will be successful in achieving high shares of sustainable modes. For Manila, the improvement of bus and train services, together with the integration of informal transit services to the structured transportation network will be critical to its success.

Although certain policies have a higher impact on transportation demand than others and are more effective in some cities than others, the influence of one policy on another is often uncertain and difficult to evaluate. A policy with a relatively smaller impact than another policy can serve a complementary or supporting role in a complex transportation system. In addition, the same set of policies and measures can also trigger different outcomes, as shown in this study. Cities are diverse in terms of their existing mode choice and transportation services, which will lead to different policy impacts, even within the same country or region.

Acknowledgments

The author thanks Claire Alanoix and Seiya Ishikawa at the International Transport Forum (ITF) for their technical assistance and the Japan International Cooperation Agency (JICA) for the household travel survey data.

Author Contributions

The author confirms sole responsibility for the following: study conception and design, data collection, analysis and interpretation of results, and manuscript preparation.

References

1. ITF. *International Transport Forum Transport Outlook 2017*. OECD Publishing, Paris, France, 2017.
2. IMF. *World Economic Outlook (WEO) Database*. International Monetary Fund, Washington, D.C., 2017.
3. OECD. *Economic Outlook for Southeast Asia, China and India 2017: Addressing Energy Challenges*. OECD Publishing, Paris, France, 2017.
4. Clean Air Asia. *Accessing Asia: Air Pollution and Greenhouse Gas Emissions Indicators for Road Transport and Electricity*. Clean Air Asia, Manila, the Philippines, 2012.
5. IEA. *Energy Technology Perspectives 2015: Mobilising Innovation To Accelerate Climate Action*. OECD Publishing, International Energy Agency, Paris, France, 2015.
6. Acharya, S., and S. Morichi. Managing Motorization and Timing of Mass Rapid Transit in East Asian Megacities. *Proceedings of the Eastern Asia Society for Transportation Studies*, Vol. 6, 2007, pp. 73–87.
7. Dissanayake, D., S. Kurauchi, T. Morikawa, and S. Ohashi. Inter-regional and Inter-temporal Analysis of Travel Behaviour for Asian Metropolitan Cities: Case Studies of Bangkok, Kuala Lumpur, Manila, and Nagoya. *Transport Policy*, Vol. 19, No. 1, 2012, pp. 36–46.
8. Sotheary, P., and M. Kunthear. Congestion Costing $6 Million a Month. *The Phnom Penh Post*, October 28, 2015. www.phnompenhpost.com/national/congestion-costing-6-million-month. Accessed October 19, 2016.
9. The World Bank. *Malaysia Economic Monitor: Transforming Urban Transport*. The World Bank, Washington, D.C., 2015.
10. Kov, M., and T. Yai. Exploring Factors Associated with Motorcyclist Accident Involvement in Phnom Penh. *Journal of the Eastern Asia Society for Transportation Studies*, Vol. 9, 2011, pp. 1796–1811.
11. CAIT. *CAIT Climate Data Explorer*. World Resources Institute, Washington, D.C.www.cait.wri.org. Accessed July 3, 2016.
12. Yagi, S., D. Nobel, and H. Kawaguchi. Time-series Comparison of Auto/Motorcycle Ownership and Joint Mode and Destination Choice Models Based on Two Large-scale Surveys in Jakarta. *Asian Transport Studies*, Vol. 3, No. 1, 2014, pp. 46–65.
13. Tuan, V. A. Mode Choice Behavior and Modal Shift to Public Transport in Developing Countries – the Case of Hanoi City. *Journal of the Eastern Asia Society for Transportation Studies*, Vol. 11, 2015, pp. 473–487.
14. Hyodo, T., C. M. Montalbo, Jr., A. Fujiwara, and S. Soehodho. Urban Travel Behavior Characteristics of 13 Cities Based on Household Interview Survey Data. *Journal of the Eastern Asia Society for Transportation Studies*, Vol. 6, 2005, pp. 23–38.
15. Choi, C. C., and N. Loh. Transport Policies and Patterns: A Comparison of Five Asian Cities. *Journeys*, September 2013, pp. 69–78.
16. Denpaiboon, C., and H. Kanegae. Compact City Strategy of Bangkok Mega-city. *Manusya: Journal of Humanities*, Vol. 11, No. 2, 2008, pp. 38–55.
17. Cervero, R. Built Environments and Mode Choice: Toward a Normative Framework. *Transportation Research Part D*, Vol. 7, 2002, pp. 265–284.
18. Manheim, M. The Demand for Transportation. In *Fundamentals of Transportation Systems Analysis* (Vol. 1: Basic Concepts, Chapter 2), MIT Press, Cambridge, MA, 1979.
19. Ben-Akiva, M., and S. R. Lerman. *Discrete Choice Analysis. Theory and Application to Travel Demand*. The MIT Press, Cambridge, MA, 1985.
20. Small, K. A., and C. Winston. The Demand for Transportation: Models and Applications. In *Essays in Transportation Economics and Policy. A Handbook in Honor of John R. Meyer* (Gomez-Ibanez, J., W. B. Tye, and C. Winston, eds.), Brookings Institution Press, Washington, D.C., 1999.
21. Small, K. A., and E. T. Verhoef. Costs. In *The Economics of Urban Transportation*, Routledge, New York, NY, 2007, pp. 56–118.

22. McFadden, D. The Measurement of Urban Travel Demand. *Journal of Public Economics*, Vol. 3, 1974, pp. 303–328.
23. Train, K. *Discrete Choice Methods with Simulation*, 2nd ed. Cambridge University Press, Cambridge, MA, 2009.
24. Koppelman, F., and V. Sethi. Closed Form Discrete Choice Models. In *Handbook of Transport Modeling* (Hensher, D. A., and K. J. Button, eds.), Pergamon Press, Oxford, UK, 2006.
25. United Nations. *World Urbanization Prospects: The 2014 Revision*. Department of Economic and Social Affairs, United Nations, New York, NY, 2014.
26. JICA. *The Comprehensive Urban Development Programme in Hanoi Capital City of the Socialist Republic of Vietnam (HAIDEP). Final Report*. Japan International Cooperation Agency, Tokyo, Japan, 2007.
27. Oort, C. J. The Evaluation of Travel Time. *Journal of Transport Economics and Policy*, Vol. 3, 1969, pp. 279–286.
28. Small, K. Fundamentals of Economic Demand Modeling: Lessons from Travel Demand Analysis. In *Decision-Based Design: Making Effective Decisions in Product and Systems Design* (Chen, W., K. Lewis, and L. C. Schmidt, eds.), ASME Press, New York, NY, 2006.
29. CIDS. *Minimum Wage Compliance Report Cambodia*. Cambodia Institute of Development Study, Phnom Penh, Cambodia, 2011.
30. Becker, G. S. A Theory of the Allocation of Time. *Economic Journal*, Vol. 75, 1965, pp. 493–517.

The Standing Committee on Transportation and Sustainability (ADD40) peer-reviewed this paper (18-04857).

Sage Publications USA
sagepub.com

ISBN 978-1-5108-7730-6

9 781510 877306

CURRAN ASSOCIATES INC.
proceedings.com